LIGHTHOUSES
OF THE
GEORGIA COAST

MUP/ H1001

© 2021 by Mercer University Press

Published by Mercer University Press
1501 Mercer University Drive
Macon, Georgia 31207
All rights reserved

9 8 7 6 5 4 3 2 1

Books published by Mercer University Press
are printed on acid-free paper that meets the
requirements of the American National Standard for Information
Sciences—Permanence of Paper for Printed Library Materials.

Printed and bound in Canada

Book design by Burt&Burt
This book is typeset in Minion Pro & Meta Pro

ISBN 978-0-88146-775-8

Cataloging-in-Publication Data is available
from the Library of Congress

LIGHTHOUSES
of the Georgia Coast

WILLIAM RAWLINGS

MERCER UNIVERSITY PRESS
MACON, GEORGIA
2021

Author's Preface   VII

Part One
Beacons of Hope

1. The Mystique of Lighthouses   3
2. Lighthouses through the Ages   13
3. Lighthouse Construction   25
4. Throwing the Light   33
5. American Lighthouses   53
6. Lighthouses during the American Civil War   69
7. Keepers of the Light   83

Part Two
Lighthouses of the Georgia Coast

8. Tybee Lighthouse   103
9. Cockspur Lighthouse   119
10. Sapelo Lighthouse   133
11. St. Simons Lighthouse   145
12. Little Cumberland Lighthouse   163

Appendix

Glossary of Lighthouse-Related Terms   183
Suggested Further Reading   188
Image Credits   190
Endnotes   193
Index   199

Acknowledgments   203

Opposite, Sapelo Island lighthouse.

AUTHOR'S PREFACE

For many years lighthouses have fascinated me, perhaps because of summers spent at St. Simons and other islands along the coast of my home state of Georgia, or perhaps because of their intrinsic beauty and the aura of mystery that has come to be associated with them. In this book, I hope to help introduce others to these intriguing seaside beacons. Technical marvels in their heyday of decades past, monuments and witnesses to historical events great and small, lighthouses have also become symbols that reach into the metaphysical, representing hope, salvation, and guides to safe harbors amidst the storms of life. And they are pleasing to the eye.

This book is not meant to be a definitive work on lighthouses. Instead, it was written for those wanting to know more about the lighthouses of the Georgia coast and for those interested in historical developments and technological advances that made these towers important navigational assets. To accomplish this, I have divided it into two parts. The first seven chapters discuss lighthouses in general, their history, technical aspects, and construction. There is a chapter outlining the development of the American system of lighthouses and other navigational aids, a chapter that focuses on the intriguing history of the lighthouses of the Confederacy, and another that examines the lives, challenges, and heroism of lighthouse keepers.

The five existing lighthouses of the Georgia coast are each discussed in detail in the five chapters that follow. Whenever possible, I have tried to include some of the fascinating stories and tidbits of history that are associated with each one. To name a few: the story of Savannah's famed "Waving Girl," the daughter and sister of lighthouse keepers; the destruction of the lighthouse towers at

Tybee and St. Simons by Confederate forces during the Civil War; the saga of the slave ship *Wanderer*, piloted into St. Andrew Sound by the keeper of the Little Cumberland lighthouse. In an appendix, I have provided a glossary of lighthouse-related terms and assembled an annotated reading list for those who wish to delve further into the subject. As many of the concepts and descriptions can be more easily understood with the assistance of photographs, diagrams, and illustrations, these are used liberally to supplement the text.

First, as a bit of background information, it is appropriate to place this narrative in its proper geographic perspective. The coast of Georgia is approximately one hundred miles in length, stretching in a northeast to southwest direction from the Savannah River in the north to Cumberland Sound and the St. Marys River in the south. The area near the Florida border is the westernmost point of America's Atlantic coast, a feature that for geologic and climatic reasons offers this part of Georgia some protection from tropical hurricanes.

Like much of the Atlantic coast from New Jersey to the southern tip of Florida, the mainland of the Georgia coast is separated from the sea by a series of so-called barrier islands interspersed among marshes and tidal creeks. These are comprised of eight major islands and/or island groups from north to south: Tybee/Little Tybee, Wassaw, Ossabaw, St. Catherines, Sapelo/Blackbeard, St. Simons/Little St. Simons/Sea Island, Jekyll, and Cumberland/Little Cumberland. Over millions of years of geologic time, the seacoast of what is now Georgia varied with climatic change, at various times located both far inland and far at sea from its present location. The Pleistocene Era, which began about 2.6 million years before the present and lasted until 11,700 years ago, was characterized by intermittent periods of global cooling and glaciation that caused sea levels to fall as much as 100 meters (330 feet) or more compared to today. This was followed by the current (Holocene) era, a time of relative global warmth and rising sea levels caused by the melting of the Pleistocene's glaciers. Geologic remnants of barrier islands formed during the Pleistocene Era have been identified as far inland as the eastern edge of the Okefenokee Swamp, some forty miles from the present coast. As they exist today, Georgia's barrier islands were for the most part formed during the late

Pleistocene and current Holocene eras. The westernmost islands, for example, Sapelo, St. Catherines, and St. Simons, date to the Pleistocene, whereas Sea Island, Tybee, and Little Cumberland are examples of those formed during the Holocene. All the islands are subject to continuous remodeling by sea currents, wind, and storms.

Although we may at times think of them otherwise, lighthouses were built as utilitarian structures, their design and location chosen to fill a need perceived at the time of their construction. All of Georgia's existing lighthouses are built on barrier islands, their primary purpose being an economic one, the guiding of ships carrying trade goods (and sometimes passengers) in and out of potentially treacherous waterways between the open sea and ports on the mainland. As such, they have become symbols of the economies of the past; in the case of Georgia, the trading of cotton, rice, timber, naval stores, and other products. In times of war, they took on strategic military significance. Today, they primarily serve as historic monuments and touristic attractions.

Although human habitation and commerce have altered in various ways the past natural status of the barrier islands, most remain relatively isolated and undisturbed. Only Tybee, St. Simons, Sea Island, and Jekyll have direct public-road access to the mainland. Others, for example, Cumberland and Sapelo islands, are open to visitors on a limited basis but are subject to restrictions on use and development. It should be appreciated that up until the early days of the twentieth century, these islands were rural wildernesses. The intense commercial development that one sees today on St. Simons and Tybee islands belies the isolation in which their lighthouses once existed.

I trust that readers will find this book both informative and enjoyable. In the process, I hope that it increases the awareness and understanding of the historical importance of Georgia's unique coastal heritage.

William Rawlings
Sandersville, Georgia
August 1, 2020

PART ONE

BEACONS OF HOPE

CHAPTER 1

THE MYSTIQUE OF LIGHTHOUSES

MELANCHOLY SHIPWRECKS

The Great Storm of early winter 1839 would be, if contemporary accounts are to be believed, the worst in living memory. It began on Saturday, December 14th, an abrupt change after "a season of unprecedented mildness and beauty," to quote one report. First there was rain and "fitful and terrific gusts" of wind, accompanied by an abrupt plunge in temperature and followed soon on by snow, whose drifts extended to the second story of houses. Chimneys were toppled, roofs blown away. New England and New York were hardest hit, with the foul weather extending as far south as Baltimore, Washington, and Philadelphia. Roads were impassable. Communities were isolated. Railways were blocked. The all-important mail—the era's link to the outside world—did not arrive. In recounting the damage of the storm, a local newspaper in a western Massachusetts town speculated, "If the storm was so severe on land, what disastrous consequences must have attended it on the ocean? We apprehend frightful disasters will fill our next paper."[1] These thoughts would prove prescient.

This storm of the last weekend of autumn was just the beginning. On each of the following two weekends, the pattern repeated itself, with hostile weather conditions extending into the first two weeks of the New Year. Inland temperatures plummeted to subzero levels. At Utica, New York, for example, the temperature on New Year's Day 1840 was 13 degrees below zero. Eighty miles or so to

Fig. 1A An illustration from an 1865 edition of "Lighthouses" by David Stevenson. Stevenson was a member of a prominent Scottish family of lighthouse builders that also included the novelist Robert Louis Stevenson.

the east, the village of Watervliet, just north of Albany, recorded a figure of 20 degrees below. Rivers iced over. Ports were clogged with sea ice.

Despite a growing manufacturing industry, the economy of the region was highly dependent on its ports and seaborne commerce. Here the true magnitude of the disastrous storm was most clearly manifest. Seafaring was a dangerous occupation; that was acknowledged. Ships lost at sea or smashed upon some coastal reef or rocky shore were common, so much so that most newspapers in port cities carried a regular column titled "Shipwrecks," or often "Melancholy Shipwrecks" if the events were accompanied by loss of life. As reports began to filter in, the extent of the storm's damage to the shipping industry became evident. In Boston Harbor, dozens of ships were blown ashore, many being sunk or severely damaged. At Cape Ann, near Rockport, Maine, some twenty vessels were driven ashore, "of which sixteen went to pieces and many of the individuals on board, probably a fourth part, were drowned," according to one newspaper.[2] The paper's informant "saw seventeen dead bodies lying on the beach; among them was the body of a woman found lashed to the windless bitts of a Castine [Maine] schooner."[1] A few miles to the south, of the sixty vessels that had sought shelter in Gloucester's outer harbor, twenty-one were driven ashore, with twenty of these being a total loss. Of those ships that remained in the harbor, all but seven had to cut their masts to survive the storm.

A summary of the loss of life and property during the Great Storm was sobering. In an appeal for aid from the American Seamen's Friend Society, it was noted that eighty vessels had been lost during the first two weeks of December 1839.[3] The storm of the 14th through the 16th of the month resulted in the total loss of eighty-nine additional vessels and "about ninety lives." The storms of the latter half of December added eighty-four more vessels with eighty-nine lives lost. The first two weeks of January 1840 saw eleven vessels and "about 125 men" appended to the list, bringing the total to 192 seafaring vessels and "about 300 lives" lost in a period of six weeks. While tragic, such loss of life was neither unusual nor unexpected. During the preceding years of 1836 through 1839, the number of American vessels lost annually due to "shipwrecks and disasters at sea" were 316, 493, 427, and 442, respectively. The number of lives "known to be lost" averaged 854 per year but were likely "much greater," as many ships were simply reported as "missing" or otherwise lost.[4,II]

Out of the myriad stories of the storm and the deaths it brought, individual tales began to emerge, most tragic, some heroic. The brig *Pocahontas*, having sailed in late October from Cadiz, Spain, bound for her homeport of Newburyport, Massachusetts, was observed dismasted and stranded on an offshore reef on Monday morning, December 23rd, following the second blow of the Great Storm. Despite being only 150 yards from the beach, the raging seas made any attempt at a rescue impossible. Initially, three men were observed clinging to the ship, one naked and lashed to

---

I "Bitts" is defined as "a pair of posts on the deck of a ship for fastening mooring lines or cables."

II While these numbers of deaths may seem relatively small in the general scheme of things, it should be appreciated that in 1840, the entire population of the United States was approximately 17.1 million, slightly more than a twentieth of the county's population 180 years later in 2020, making the proportional impact of these losses far greater.

the taffrail, the other two clinging to the bowsprit. Within a short while, the sea had swallowed two of the men, leaving only one on the bowsprit. As the crowd that had gathered on the beach watched in horror for several hours with no means to rescue him, he was swept away, leaving no survivors. It was theorized that the *Pocahontas* had anchored offshore for the night, but the force of the gale had thrown her against the rocks, resulting in death of all on board. The ship's name and that of her captain, James G. Cook of Newburyport, were discovered from debris found on the beach. The bodies of the crew of the vessel were lost to the sea, their names and homeports unknown. One report commented,

When she came into the bay, and whether those on board knew her position during the gale; whether the majority of them were swept over together, or one by one, being overpowered by the intensity of the cold and the violence of the sea, will never be known as not one of the twelve or thirteen souls on board is left to tell the sad tale. It is heart-rending indeed, that the toil-worn mariner, after beating about on a stormy coast for many days, should be wrecked and perish within sight of the smoke ascending from his own hearth.[5]

There was another shipwreck the preceding weekend, yet another story of life and death, but this time with a somewhat more positive outcome.[6] The schooner *Deposit* out of Belfast, Maine, with a crew of seven, was forced ashore in Ipswich Bay by the storm near midnight on Sunday, December 15th. A man named Marshall discovered the wreck the next morning and alerted a Mr. Greenwood, described as "Keeper of the Lighthouse." The vessel was close to shore, but the wind-driven waves made it impossible to reach by boat. Greenwood waded into the sea with one end of a rope, fighting his way to the stranded vessel. Marshall tied the other end to a small boat, which Greenwood then pulled through the surf to the *Deposit*'s side. As waves broke over the ship's deck, they discovered the captain, a man named Cotterall, "almost senseless and completely exhausted." His wife was "exhausted" as well but in somewhat better condition. One crewman was dead, another *in extremis*. Greenwood assisted Marshall in lowering the captain into the small boat, but it capsized almost immediately, throwing the latter two into the sea. The captain disappeared in the waves, while Marshall saved himself by grabbing on to a rope. Taking hold of the captain's wife, Marshall and Greenwood rode a wave to shore, thus saving her life. Two of the crew had managed to make it to shore on their own; four others, including the captain, drowned. As the lighthouse keeper later recalled, "The horrors of the storm, the sight of the dead around him, and the cries of the dying for succor were nothing to the terrific shrieks of the captain's wife as she saw her husband buried beneath the waters." In spite of the danger, though, as a lighthouse keeper he was willing to risk his life to save the lives of others whom he'd never met.

The reactions to the Great Storm ran the gamut from sorrowful to resigned. There were tragedies to be sure, but those who chose to challenge the sea often lost the match. No matter what the emotions, there were the issues of commerce and trade, which took the vessels to sea in the first place. With regularity, newspaper accounts listed the cargoes and the amounts for which they and the ships were insured. The bark *Lloyd*, which foundered on Maine's Nantasket Beach on December 23rd, was returning from Havana with a cargo of 550 hogsheads of molasses, 80 thousand cigars, 5 mahogany logs, 11 bags of coffee, 4 casks of pumpkins, 10 pipes of aguardiente, 40 casks of wine, 56 jars of olive oil, 5 bags of aniseed, a quantity of oranges, and one box of sugar. The cargo was insured for twelve thousand dollars, the vessel for six thousand. The cargo, strewn along the beach, was immediately placed under the charge of Robert Gould, Esq., the commissioner of Wrecks. Such events were common enough to warrant a governmental office whose purpose was to see to the salvaged goods' proper disposition. Of the crew of ten, for the most part nameless, only one, an Englishman named George Stott, survived.

The Great Storm of the winter of 1839–1840 was an event that should have reasonably been expected. Could more lives have been saved? In its wake, an editorial in the Boston *Gazette* opined,

There is not a Life Boat of proper construction, so far as we know, on our immediate coast where so many lives are annually lost. In the late unfortunate shipwrecks in our bay, very many lives might have been saved by a single life boat. At Gloucester, for example, persons were seen in the wrecks and on the rigging of the vessels during the night of the storm until swept off or immersed and drowned when there was no possible means of saving them at hand. We have seen two or three of the men who were saved from the wrecks during that dreadful scene of horror and death, and from their stories we are assured that with a Life Boat, well-manned by the hardy and bold people of Gloucester, many of the persons drowned there might have been saved. So also in the case of the barque *Lloyd*—a lifeboat would have saved her whole crew, and perhaps the vessel. . . . It is a fact, which does not speak well for our humanity, that our rocky and dangerous coast is left deficient of *any means* of rendering assistance to distressed or wrecked vessels. Interest, as well as humanity, would seem to call loudly upon this community to supply so great a want.[7]

WHY LIGHTHOUSES?

Oftentimes we assume things to be obvious, perhaps because that is the way we perceive it has always been. If asked about the purpose of lighthouses, most would say that they exist to serve as daymarks[III] by day and beacons at night to prevent the needless loss of human life. Yet a more realistic look at the history of such structures over the past few centuries might suggest otherwise.

For the vast majority of recorded history, humans have lacked the ability to predict such perilous maritime events as hurricanes or other dangerous storms. Even today, while we can make preparations to avoid their damage, we can do nothing to prevent them. As such, these are often referred to as "acts of God." The term, and its implications, has firmly entrenched itself in the canons of Western law. For example, a mid-nineteenth century case heard before a Scottish appeals court defined such events as "ocurrencies and circumstance which no human foresight can provide against, and of which human prudence is not bound to recognize the possibility; and which, when they do occur, therefore, are calamities that do not involve the obligation of paying for the consequences that may

---

[III] See glossary for definition of "daymark" and other words or terms related to lighthouses that may not be familiar to the average reader.

result from them."[8] Granted, a life at sea was one of danger and potential death as New England's Great Storm so amply demonstrated. Anyone who chose such a career was well aware of, and presumably willing to face, the consequences. If God chose to send a mighty storm that cast sailors into a raging sea, they knew and had accepted the risk; it was His Will.

A closer look at the timeline of lighthouse history reveals an important correlation: the growth in their number roughly parallels the growth in seaborne trade. While the loss of life at sea was tragic and heartrending, transportation of the ships' cargoes was the purpose of the voyages. It was often ships' owners, investors, and insurance syndicates who were the first to call for the construction of a lighthouse in a particularly dangerous spot. More than a few early lighthouses were private endeavors, funded not by a governmental authority, but rather by tolls on passing vessels and their cargoes. Trinity House, the organization that became the British lighthouse authority, was originally established as a private corporation in 1514 under the terms of a royal charter issued by Henry VIII. In 1696, for example, the corporation signed an agreement with Henry Winstanley to construct a lighthouse on the infamous Eddystone Rocks, on the approach to the port of Plymouth in southwest England.[IV] The project was to be funded by ship levies, all of which Winstanley was to receive for the first five years, and then share equally with Trinity House for the next half century. As time passed, and the importance of maritime trade to national economies became better appreciated, most lighthouses were erected under the direct or indirect auspices of states and/or nations. As a result, a greater emphasis was placed on the importance of human life such that by the nineteenth century this became, in popular imagination, the prime reason for the existence of these structures.

[IV] See Chapter 2 for more detailed information on the Eddystone lighthouses.

THE LIGHTHOUSE AS A SYMBOL

Over the centuries the lighthouse has become more than a physical object. For many, it is seen as a powerful symbol of hope, a metaphysical beacon of guidance for those tossed about on the stormy seas of life, a psychological landmark pointing the way to a place of shelter. Likewise, lighthouse keepers are often romanticized, portrayed as saintly men and women willing to forsake a bountiful life in the city for the reclusive existence of a lighthouse, selflessly dedicating their lives to the salvation of others.

Lighthouses' locations on remote coasts or amidst stormy seas have imbued these towers with the concepts of isolation, mystery, and romance. This is reflected in the many literary works that incorporate lighthouses into their plots or settings. Seemingly innumerable academic papers have been written about the symbology of Virginia Woolf's *To the Lighthouse*. Lighthouse venues have become a staple of literary works of mystery, ranging from Jules Verne's posthumously published adventure tale, *The Lighthouse at the End of the World* (1906), to P. D. James's crime thriller, *The Lighthouse* (2006). Robert Louis Stevenson, a member of the famous Stevenson family of lighthouse builders and originally apprenticed to that profession, incorporated lighthouse themes in several of his works. Edgar Allen Poe, the great American author, was working on a lighthouse-themed story at the time of his death in 1845. Eugenia Price's *Lighthouse* (1972) is a work of historical fiction set on St. Simons Island, Georgia. And lest those who manned lighthouses be denied, *The Lighthouse Keeper's Daughter* (Hazel Gaynor, 2018) is a historical novel whose plot is centered about the families of lighthouse keepers. Taking the association to the extreme, the fictional DC Comics superhero Aquaman, in one iteration of his backstory, is the product of the mating between a lighthouse keeper and a water-breathing refugee from the underwater Kingdom of Atlantis.

On a more mundane level, the Book Cover Designer, a digital marketing company selling ready-made book covers to publishers and would-be authors, offers dozens of colorful designs featuring lighthouse themes.[9] And as a symbol of romance, many a cover of romance fiction features amorous couples entangled on seaside beaches with the image of a lighthouse standing erect in the distance.

An online search reveals lighthouse-themed tattoos, lighthouse jewelry, and instructions for interpreting the significance of lighthouses perceived in tea leaves. Among the many and diverse associations between enterprise and lighthouses is the logo of Safe Passage Urns, a supplier of cremation urns. This company chose the symbol of a lighthouse as its logo, noting,

> Lighthouses have long been used in literature and cinema to symbolize strength, safety, individuality, and even death. Because lighthouses are constructed to withstand powerful storms and turbulent ocean waters, it is no wonder why they are frequently depicted as symbols of strength. Constructed on the top of hills or cliffs, they have a resilient nature. They represent hope and safe haven when a ship first sees the glimmering light atop a lighthouse in the distance. Much as the calm waters of a harbor embody a final resting place for a sea-weary ship, our cremation urns also represent the final resting place of a loved one's ashes. At Safe Passage Urns, our goal is to be your guiding light for funeral planning, estate distribution, and grief support. We aim to be as reliable as the lighthouse light that shines brightly.[10]

The perceived selfless nature of those who tend lighthouses is reflected, for example, in The Lighthouse for the Blind, Inc., an incongruously named private nonprofit corporation founded in 1918 whose initial purpose was (in part) to assist soldiers blinded in the battles of World War I.[11]

BEAUTIFUL ANACHRONISMS

Twenty-first century reality dictates that the day of the lighthouse is long past. Advances in science, technology, and navigation have reduced these majestic towers to vestiges of a different era when a career at sea was undertaken only by the adventurous or foolhardy. Yet their beauty continues to enthrall us, while the tales of lighthouse lore that have accumulated over the centuries continue to intrigue and inspire us. Beauty, lore, and inspiration aside, to fully appreciate lighthouses and their mystique, one must understand the basics of their history, construction, and function. The chapters that follow address these topics. While at times technical and even arcane, a good working knowledge of the hows and whys of lighthouses will vastly increase one's appreciation of these beautiful anachronisms.

CHAPTER 2

LIGHTHOUSES THROUGH THE AGES

The lure of the sea has attracted man since the dawn of human civilization. Perhaps it was the search for new lands to settle, new trading partners, or simply human curiosity that drove prehistoric peoples to explore the unknown and often dangerous world beyond the horizon. There is abundant archeological evidence of seafaring craft prior to 6000 BC. In the eastern Mediterranean, Egyptian vessels plied trade routes as early as the First Dynasty (c. 3150–c. 2890 BC). More than a thousand years before the birth of Christ, the Phoenicians had established an extensive trading network in the Mediterranean basin, by some accounts even reaching Europe's Atlantic coasts and the British Isles. In the Far East, Polynesian explorers took to the sea to populate the islands of Southeast Asia between 3000 and 1000 BC. By shortly after 1000 AD, the descendants of these Pacific Islanders had reached Rapa Nui (Easter Island) in the South Pacific and the islands of Hawaii to the north, having navigated their canoes over as much as two thousand miles of open ocean.

During the centuries of Rome's dominance in the West, the Mediterranean, then known as the *Mare Nostrum*, was the heart of a thriving flow of international trade in goods from all the known world. Prior to the invention and general use of precision navigational instruments, most European ships engaged in commerce and trade sailed within sight of land. The magnetic compass, invented during the Han Dynasty in China about 200 BC, did not come into general use as a navigational aid in Europe until the twelfth century AD. For those sailors brave enough to venture out of sight of land, the positions of the sun and stars provided guidance, as did the currents, the winds, and the observation of birds. Mediterranean seafaring was mostly seasonal. Weather conditions between the months of April and October tended to be calm and predictable, and storms infrequent. Vessels hugging the often-rocky shoreline were reasonably safe from danger during daylight hours. In contrast, the late fall, winter, and early spring months could bring violent storms and unpredictable weather, leading to marked restriction of seaborne trade.

The decline of the Roman Empire near the end of the fifth century AD ushered in the Middle Ages in Europe, a thousand-year period characterized by the collapse of centralized authority, the depopulation of once-mighty cities, and threats to social stability as various tribal, religious, political, and military interests vied for power. The fifteenth century heralded the beginning of the so-called Age of Discovery, a period of European exploration beyond the world of the Mediterranean. Portuguese explorers charted a sea route to India. Columbus's voyages for Ferdinand and Isabella led to the "discovery" of the Americas and, in turn, wealth that would make Spain a major European power. French, English, and Dutch voyages of exploration would soon follow. While well-established land trade routes between Europe and Asia had existed for centuries, within a matter of decades new thriving routes of sea trade developed, accompanied by European colonization in Africa, Asia, and the Americas.

ANCIENT LIGHTHOUSES

Early lighthouses in the Mediterranean region served several purposes. In addition to their function as navigational landmarks, they were used as observation towers to warn of impending danger and for signaling both to ships and inland. By the time of the Roman Empire, dozens of such towers were located at strategic points around the Mediterranean and Black seas, most often to assist in guiding ships into the harbors of major trading centers. Only in rare cases do their ruins still exist today. Perhaps the best

Fig. 2A The Pharos of Alexandria.

known was the fabled Pharos of Alexandria, one of the Seven Wonders of the Ancient World. The lighthouse, located at the entrance to the harbor of Alexandria, was completed in the year 280 BC by Ptolemy II Philadelphus. It was built on Pharos Island, from which the structure took its name. Importantly, the name of the island became synonymous with lighthouse in the Romance languages, for example, *phare* in French, *faro* in Spanish and Italian, and *far* in Romanian and Albanian. The study of lighthouses is known as *pharology*.

Alexandria's lighthouse was, by all historical accounts, a truly massive structure constructed of hewn stone. Although scholars disagree on its exact height, most estimates place it between 100 and 120 meters (approximately 330 to 395 feet), with a base of 30 meters (approximately 100 feet) on each side.[1] The lighthouse was constructed in three sections: it had a square base topped by an octagonal tower, which was, in turn, topped by a round tower. A platform for a signal fire to guide ships into the harbor crowned the apex. The exterior was said to have been richly adorned with

---

[1] Some early sources describe the Pharos as even taller than Giza's Great Pyramid, whose height originally measured approximately 147 meters (about 480 feet).

statuary, with the top accessible by an interior ramp. Although some writers said the light could be seen at very great distances, the geographic range of a tower 350 feet tall would be about twenty-two nautical miles. Over the ensuing centuries, the lighthouse fell into disuse, damaged by a series of earthquakes. In the late fifteenth century, Pharos Island became the site of a harbor fort built on the ruins of the lighthouse by Qaitbay, then sultan of Egypt.

Another well-known lighthouse of the era was the Pharos of Ostia, located at the harbor of ancient Rome near the mouth of the River Tiber. As Rome's power and economy depended as much on trade as it did on military might, the port of Ostia was reputed to be one of the finest in the empire. Emperor Claudius rebuilt the harbor during the middle of the first century AD. Two long breakwaters guarded the entrance to the harbor, between which was an island serving as the base for Claudius's lighthouse. Although no exact descriptions of it survive, it was said to be modeled on the Pharos of Alexandria, though smaller in size. Contemporary mosaic depictions of the lighthouse show it to be three stories in height, with a cupula on the summit where a signal fire provided light. A large statue of the emperor at its base greeted incoming sailors. In addition to serving as a harbor light, the structure provided an important platform for the defense of the harbor in case of attack. With the decline of Rome, the lighthouse fell into disrepair, although remnants of it were said to still be visible in the sixteenth century. Due to silting of the river delta, the area of the original Roman port is today several kilometers inland from the coast.

One of the few currently remaining Roman-era lighthouses is found on a promontory above the port of Dover, in southeast England.[11] It dates from the first century, built after Claudius's successful invasion of Britain in 43 AD. Located only about eighteen nautical miles across the English Channel from the coast of France, Dover became a major sea and trading port during the nearly four centuries of Roman rule in Britannia. There were origi-

---

11 The *Farum Brigantium*, or Tower of Hercules, is another Roman lighthouse dating from the first century AD and located near the entrance of La Coruña harbor in northwest Spain. It was extensively rebuilt and "augmented" in the eighteenth century by the architect Eustaquio Giannini, in the process losing and/or obscuring many of its original features.

nally two towers near Dover, one on either side of the port area, as well as a third tower located across the channel in France at what is now the city of Boulogne-sur-Mer. It was said that in cooperative weather each tower could be seen from the others, making the small network an asset for navigation and for signaling in times of danger. Of the three towers, only Dover's easternmost survives partially intact.

The Dover lighthouse was built of stone and Roman brick. It originally stood approximately eighty feet in height and may have had as many as eight stories. An open fire burned on its summit. Its survival through the years is likely due to its location on the grounds of Dover Castle, where it would have served as a lookout tower during the turbulent years of the Middle Ages. By the thirteenth and fourteenth centuries it had fallen into disrepair. During that era the topmost portion was rebuilt with the addition of battlements, and the structure repurposed as a church bell tower. Today the former lighthouse measures about sixty feet in height and stands adjacent to the Church of St. Mary de Castro, dating from about 1000 AD.

Fig. 2B The Roman lighthouse at Dover.

THE SHRINKING WORLD IN THE AGE OF DISCOVERY

The European voyages of exploration that began in the fifteenth century brought with them new challenges in the science of navigation. As transoceanic journeys became more common and lucrative trade routes developed, travel into hitherto unknown parts of the world required more accurate ways of determining a vessel's location. A sailing ship crossing the Atlantic Ocean, for example, could well be out of sight of any earthly points of reference for weeks or months. The relationship of the earth to the sun, stars, and planets had been reasonably well known since ancient times. In this new age of exploration, however, the sciences of astronomy, navigation, and mapmaking took on greater importance. The concepts of latitude, measured by concentric circles north or south of the equator termed *parallels*, and longitude, marked by *meridians* running through the North and South poles, were well established as early as the second century AD. Defining one's latitude could be accomplished through the use of an astrolabe (or one of its predecessors), an instrument used to determine the position of the observer in reference to known celestial bodies. Longitude, however, presented a thornier problem.

Unlike the parallels of latitude, which are fixed in their relationship to the equator and equidistant from one another, the meridians of longitude intersect at the North and South poles and have no natural reference point. At the equator, the circumference of the earth is approximately twenty-five thousand miles. For the 360-degree circle represented by the equator, a fixed point on this imaginary line makes a full rotation every twenty-four hours. During each hour this point moves fifteen degrees, or just over 1,000 miles. Stated another way, one degree of the earth's motion at the equator represents a distance of about seventy miles. However, a similar point on a meridian near the North or South geographic pole would move only a few miles during an hour of the earth's rotation.

For navigational purposes, one degree of the earth's rotation represents four minutes of time.[III] If a ship's captain could know the exact time at a reference location and could determine the exact time at his current position, the difference between the two would represent the ship's position in degrees east or west of the reference location. Horologically based concepts to determine longitude had been suggested in the sixteenth century and tested—without success—in the seventeenth. While the solution to the problem might appear simple, there was one major problem: even as late as the early eighteenth century, timepieces with sufficient accuracy to be of use simply did not exist.

[III] The earth's equator is assumed to be a circle containing 360 degrees. The earth completes one rotation every twenty-four hours, or, stated in minutes, every 1,440 minutes. 1,440 minutes divided by 360 degrees equals four minutes per degree.

A number of noted scientists proposed solutions to the problem of longitude based on the positions of heavenly bodies. Around 1500, Amerigo Vespucci suggested using the relative positions of the moon and Mars to estimate longitude. Shortly thereafter, German astronomer Johannes Werner proposed a somewhat similar method, relying on the position of the moon and certain stars. Nearly a century later, Galileo Galilei observed that eclipses of the moons of Jupiter could be used to determine longitude. Other noted astronomers and scientists, including Giovanni Cassini, Christiaan Huygens, Edmond Halley, and Sir Isaac Newton, proposed various astronomically based ideas.

While the problem of location at sea might seem far removed from the subject of lighthouses, through much of the seventeenth century and well into the eighteenth, the growth of seaborne commerce, colonization, and military fleets outpaced the erection of light beacons on the often-treacherous coasts of Europe. In England, only one functioning coastal lighthouse was known to exist before the year 1600. By about 1700, as many as three hundred ships a year were engaged in active trade between the British Isles and the West Indies, but the number of lighthouses had increased to only fourteen, all but two on the east coast facing the English Channel and points to the north.[1] An error of even a few miles for a returning merchant vessel could mean the difference between fortune and disaster, not the mention the needless loss of life.[IV] It was, in the end, a British naval disaster and its consequences that helped lead to a practical and reliable method of determining longitude.

On September 29, 1707, a group of twenty-one British warships left the port of Gibraltar, returning to their base in Ports-

[IV] To quote one author on the dangers of life at sea in Britain at the beginning of the eighteenth century, "Almost a third of British seamen died pursuing their trade, either killed by the punishment of life on board ship or sacrificed to storms and drownings. Nearly everything the modern mariner relies on—competent maps, accurate instruments and adequate communication—was either unreliable or nonexistent. The major sea lanes around Britain were crowded, and collisions were frequent. What is now fixed and understood was then debatable, and navigation was more of an art than science. Sailors depended on experience or luck to avoid danger, and when they did run into trouble, there was no kindly lifeboat service to save them." (Bathhurst, *The Lighthouse Stevensons*, 2)

mouth after a series of naval engagements in the Mediterranean during the War of Spanish Succession. The commander in chief of British fleets, Sir Cloudesley Shovell, led the fleet. The weather was said to be extremely bad, with constant rain, wind, and limited visibility. While the ships' navigators believed they were on course for Portsmouth, they were, in fact, more than two hundred miles to the west and some thirty miles off the Cornwall peninsula. During the night of October 22nd, Shovell's flagship struck rocks off the Isles of Scilly, sinking within three to four minutes with her entire crew of about eight hundred sailors. Three other ships sailing nearby were unable to change course and rapidly sank after running aground. The remaining ships managed to avoid the rocks, eventually making it back to their homeport. In all, between 1,400 and 2,000 seamen drowned, including Admiral Shovell. The cause of the disaster, one of the worst in British maritime history, was attributed to foul weather and the inability to determine the fleet's longitude. Ironically, of the two lighthouses on the south coast of England, one was located on the Isles of Scilly, erected in the 1680s.

Based in part on the Scilly disaster, and in response to demands from both merchants and ships' captains, on July 8, 1714, the British Parliament passed the Longitude Act, offering significant monetary rewards (equivalent to millions of dollars today) for the invention of a simple and practical method to accurately measure longitude. The level of the award was to be based on accuracy, with the top prize of twenty thousand pounds given for a method whose precision was within one-half degree.[v] Numerous solutions were suggested, including augmentation of the then-prevalent lunar method based on the position of the moon and certain stars. By 1735, after more than twenty years had elapsed, the prize had not been claimed. In the thoughts of many, the puzzle seemed unsolvable, something only a madman might pursue. That year, William Hogarth, the London-based artist and printmaker, depicted an inmate of the city's insane asylum, Bethlehem Hospital, scribbling

---

[v] The original Longitude Act of 1714 was amended and broadened several times over the next century. The Board of Longitude, established under the Act, was disbanded in 1828.

on the walls of his prison, seeking a solution to the problem of longitude.[VI]

In 1735, a self-educated Lincolnshire clockmaker named John Harrison submitted to the Board of Longitude a remarkably accurate working model of a clock designed to fulfill the criteria for use as a navigational instrument. A talented and meticulous man who always seemed to believe he could improve on his own designs, over the next decades Harrison submitted three additional and more sophisticated timepieces designed to meet the criteria specified under the Longitude Act. Hindered by a board influenced

Fig. 2C Detail from a 1735 print by William Hogarth, one of a series from "A Rake's Progress." This final print of the series shows the interior of an insane asylum. The man in the center behind the open door is scribbling on the wall the word "longitude," implying that only a madman would attempt to find a solution to this seemingly unsolvable problem.

VI The print referenced is Number 8 in Hogarth's "A Rake's Progress," produced in print form in 1735. The term "bedlam," meaning a scene of uproar and confusion, is a late Middle English word derived from the name of London's Bethlehem Hospital, which served primarily as a hospital for the insane from about the year 1400.

by politics and reluctant to acknowledge his success, he was finally awarded the Longitude Prize in 1773 only after the intervention of King George III on his behalf. The ensuing years saw the development of the mass production of highly accurate chronometers designed for the rigors of sea voyages. The Board of Longitude, in its final years, was primarily responsible for vetting the accuracy of marine timepieces for the British navy.[VII]

THE GROWTH OF TRANSATLANTIC TRADE AND THE PROBLEM OF NAVIGATION

By the end of the sixteenth century, voyages of exploration had given way to voyages of trade and plunder. Spain's conquests in Central and South America yielded sources of silver and other riches that would support the Spanish monarchy from the late sixteenth century until the dawn of the nineteenth. Britain's colonies in North America and the Caribbean were exploited for raw materials, including furs, timber, molasses, and fish. In order to provide labor in the New World, active and growing seaborne commerce in human commodities—slaves—developed between Africa and the Americas. A pattern of "triangular trade" developed: raw materials from the Americas were shipped to Europe, where they were sold or traded for manufactured goods. These, in turn, were transported to the west coast of Africa and used to purchase slaves. The slaves, in turn, were taken back across the Atlantic where their sale provided funds to purchase more raw materials. This trade pattern peaked in the eighteenth century. The United States' and Britain's ban of the African slave trade in 1808 coincided with a massive increase in the export of cotton from the American South to Europe, made possible by Eli Whitney's cotton gin in the preceding decade. In a period of less than three centuries, the vast Atlantic, for thousands of years the realm of the unknown, had become a busy highway of international trade.

VII For a detailed and in-depth review of the history of the longitude problem and its eventual solution, the interested reader is directed to Dava Sobel's *Longitude* (Walker Publishing Company, 1995). Simon Winchester's *Precision* (Harper, 2018) has an excellent section on John Harrison and his timepieces.

The increase in maritime trade was naturally accompanied by an increase in shipping lost at sea. Most wrecks occurred near seacoasts, the common factors being storms, fog and darkness, and errors in navigation. The number of recorded shipwrecks was astounding by twenty-first-century standards. For example, the seacoasts of Britain were among the best lighted in the world around the end of the eighteenth century. Between 1793 and 1829, an average of 550 wrecks occurred annually, increasing to 800 in 1833.[2] The increase in number of wrecks was attributed not so much to poor seamanship, but rather to a larger volume of shipping. Despite the losses, the number of lighthouses and other seamarks scarcely kept pace. An estimate of the number of lighthouses and lightships[VIII] "of value for coastal navigation" in selected European countries and North America between the years 1600 and 1819 is shown in the table below:[3]

EUROPEAN AND AMERICAN LIGHTHOUSES, 1600–1819

	1600	1700	1800	1819
SWEDEN	2	5	11	12
FRANCE	1	5	16	17
ENGLAND	1	14	30	37
SCOTLAND	0	1	9	15
IRELAND	0	3	8	17
UNITED STATES	—	—	24	50

Compounding the problem of the limited number of lighthouses was the fact that prior to the first quarter of the nineteenth century, those that existed were relatively limited in their luminosity and visual range. As will be noted in the following chapters, that situation would improve dramatically—at least for European

---

[VIII] Lightships, now largely obsolete, acted as water-based lighthouses in areas where traditional lighthouse construction was not possible due to depth or other factors. A lightship would be anchored in a fixed position with a lighted beacon positioned on a raised mast, as well as a foghorn or other sound-generating device. The last American lightship was decommissioned in 1985. In many situations they were replaced by lighted buoys.

lighthouses—with the breakthroughs in lighting technology discovered and perfected by such men as Ami Argand, Augustin Fresnel, and François Arago.

THE GOLDEN AGE OF LIGHTHOUSES

The nineteenth century would become the golden age of lighthouses. It was a period that began with a pressing need for coastal directional seamarks, was characterized by great leaps in the science of navigation, and whose final decades presaged the technological advances of the next century that would eventually transform these beacons of hope into relics of the past. It was a century that saw increasing transatlantic trade and the coming-of-age of the new republic of the United States of America. During these decades there was a near-exponential increase in the number of lighthouses worldwide; in the United States, for example, the number increased by more than a hundredfold between 1800 and 1900.

In order to acquaint the reader with the practical details of lighthouses, the chapters that immediately follow will address lighthouse construction and the technology behind the ever-more-powerful light sources that served as beacons. As this book focuses on lighthouses of the Georgia coast, subsequent chapters will survey the history of American lighthouses from the first constructed at the entrance to the Boston harbor in 1716 to the eventual decommissioning in recent decades of essentially all of the nation's lighthouses.

CHAPTER 3

LIGHTHOUSE CONSTRUCTION

Mention of the word "lighthouse" evokes a vision of a tall, slender structure topped by a glass-enclosed canopy that surrounds a powerful beacon casting its light out to sea. Indeed, that image could well describe the majority of such structures that exist today. But this has not always been the case. Like most forms of utilitarian construction, lighthouses have evolved with technology and changing times. While Rome's Coliseum of the first century AD and a modern sports stadium both serve the same basic purpose, they are, in almost every aspect, far apart in both form and function. Like stadiums, religious buildings, marketplaces, and homes, modern lighthouses bear only a superficial resemblance to their historical predecessors.

It should be appreciated that lighthouses in their various iterations over the centuries were empiric structures. They were built to fill a specific local need, using available material, and generally by men who, while they may have had experience as masons, carpenters, bridge builders, or laborers, had no knowledge of exactly how a lighthouse should look or even function. Most lighthouses initially served the purpose of guiding ships into harbors. While the Egyptian, Greek, and Roman civilizations (to name a few) produced architectural marvels that have withstood the ravages of time, the formal science of engineering is barely a few centuries old, having become a standard discipline in European universities only in the eighteenth century.

Relatively modern lighthouses, those built during the late eighteenth and nineteenth centuries, tend to follow a common plan of construction, with variations and adjustments based on terrain and locality. Broadly speaking, they can be divided into two types, those built on coastal land and wave-swept towers built at sea. The former commonly serve as seamarks to assist sailors in determining their position and often to warn of dangerous shores and currents. The primary purpose of the latter is the marking of dangerous rocks, shoals, sandbars, or the like that otherwise might not be recognized by a shipboard sentinel. While both are at risk for damage and destruction from gales and storm-driven waves, this factor becomes a primary consideration in the design and construction of wave-swept towers subject to the relentless pounding of the sea.

It should also be recalled that prior to the mid-twentieth century, most lighthouses required keepers, men and women whose mission was not only to maintain the light, but often to attend to fog signals, range lights, and other associated structures. For land-based lighthouses, the keepers' quarters were most often separate, with facilities sometimes including garden plots, limited grazing land for livestock, or other means of support. At the other extreme were wave-swept towers, whose keepers were generally required to live in cramped quarters inside the lighthouse itself.

Lighthouses, with a few notable exceptions, were rarely constructed of wood. While such structures built on land may take any number of forms, wave-swept and seaside towers tend to fall into one of four types, or a combination thereof: many, especially early modern lighthouses, were primarily built of masonry (stone, brick, and/or concrete) components. In areas with sandy or shifting sea floors, a second type was commonly used: an openwork steel or iron-framed support structure anchored by straight or screw-pile foundations served as a base, with enclosed compartments and a lightroom elevated far above the ocean's surface. In the United States, such types are often found in the shallow waters of the Chesapeake Bay area and along Florida's Atlantic and Gulf coasts. (See Figure 3A.) A third type, preformed cast iron lighthouses, valued for their resistance to corrosion and ease of construction, were popular in the latter half of the nineteenth century. The firm of Gustave Eiffel, for example, best known for his eponymous Parisian tower, produced dozens of such lighthouses that were erected in numerous sites around the world. Finally, wave-swept towers, built

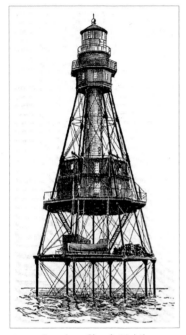

Fig. 3A American Shoals Lighthouse in the Florida Keys, circa 1880.

Fig. 3B The Cordouan Lighthouse, placed into service in 1611.

in areas without a solid foundation (e.g., near sandbars or coral reefs), were often constructed on cribs or caissons, large cylinders filled with concrete and/or stone sunk in relatively shallow waters in order to form a solid construction base.

Early "lighthouses" were often simply bonfires built on seaside promontories or lights hung in existing elevated structures such as church towers. Beginning with the increase in European maritime trade in the sixteenth century and accompanied by the often-staggering toll of shipwrecks and lost lives, efforts were undertaken to provide navigational guidance for some of the more dangerous coastal approaches. One of the earliest wave-swept lighthouses is the Cordouan light, built on a small island at the entrance of the Gironde estuary on France's southwest coast. The island had been the site of earlier fire-lighted towers since at least the ninth century AD. With increasing sea traffic, a new tower was begun in 1584 and completed in 1611. A grandiloquent structure typical of French public architecture of the day, the lighthouse was built on a fifty-foot diameter base elevating the lower part of the tower above the waves. The light, provided by a wood-burning brazier, was some 160 feet above the sea. Keepers' quarters and an elegant King's Apartment—should he ever choose to visit—were within the tower. In the late 1700s, an additional thirty-six feet in height was added to the tower's height and a new set of Argand lamps and reflectors were installed. The basic structure, declared a French national monument in 1862, still stands today.

The first British wave-swept light was built on the infamous Eddystone Rocks, some dozen miles offshore on the sea approach to the port of Plymouth. As one lighthouse historian described the area,

It is probable that about 1700, of all the underwater dangers throughout the world, the Eddystone reef shared with the extensive sandbanks off the southeast coast of England the terrible record of wrecking ships most frequently. Even at low water during daylight, little of the Eddystone reef appeared above the sea surface. Hence it formed a nearly invisible trap, dangerous particularly to vessels entering the English Channel from the west, not sure of their position.[1]

The first in a series of Eddystone lighthouses was built by Henry Winstanley, a London entrepreneur with no formal training in engineering. Eighty feet high and constructed of wood, it was completed in 1698 and illuminated with candles. After barely surviving the first storm season, Winstanley reinforced his design by adding stone cladding over the basic timber-frame construction. Foolishly confident of his tower's ability to repel the forces of nature, he and five other men elected to ride out the Great Storm of 1703 in the lighthouse. When the storm cleared, the rock had been swept clean; Winstanley and his men were never found.

A second lighthouse was completed on the Eddystone reef in 1709. This, too, was made of oaken timbers surrounding a core of stone. This structure lasted until 1755, when a spark from a lantern caused a fire that destroyed it. The third Eddystone light was built between 1756 and 1759 by John Smeaton, a civil engineer. About sixty feet in height, the base of the tower was constructed of cut stone locked together through the use of dovetail joints and sealed with "hydraulic lime," a type of concrete capable of hardening under water. For more than a hundred years, Smeaton's light warned sailors of the dangers of the Eddystone Rocks. By the 1870s, this third tower had begun to show signs of instability due to the sea's undermining of its foundation. A new lighthouse was commissioned.

Between 1879 and 1882, the fourth Eddystone lighthouse was built under the direction of James N. Douglass, the engineer in chief of Trinity House, the British lighthouse authority. By this time, the art of lighthouse building had become a science based on extensive worldwide experience. Like Smeaton, Douglass relied on dovetailed stone construction but relocated the site of the tower to a more secure portion of the rock about forty yards away. Initially equipped with Fresnel lenses and powerful concentric-wick oil lamps,[I] the tower stood 160 feet above the waves. For his service, Douglass received a knighthood.

The Bishop Rock lighthouse (Figure 3C), completed in 1881, provides a more detailed example of wave-swept lighthouse construction. This tiny islet of Bishop Rock, a bit of stony reef measuring roughly 150 by 50 feet in size, is the westernmost landfall of Britain's Scilly Islands, located less than two miles to the northwest of the Gilstone Reef, the rocks that took the lives of hundreds of Admiral Shovell's sailors in the 1707 disaster.[II] The entire area is one of hidden rocks and reefs, notorious as a mariners' graveyard. Had there been a functional lighthouse there at the time, it might well have alerted Shovell's fleet to the danger.

Construction of the first Bishop Rock lighthouse, an openwork tower of cast and wrought iron, began in 1847 but was destroyed by a storm in 1850 before its completion. A second, taller, and far more substantial tower of granite blocks was begun in 1851 and became functional in 1858. Shortly after being put into service, the tower's fog bell, weighing more than five hundred pounds and located a hundred feet above the highwater mark, was washed away during a particularly violent storm. Moreover, the tower was found to vibrate during such storms. In the 1870s attempts were made to strengthen the structure by affixing iron reinforcing ties to its interior walls. In 1881, when further evidence of damage to the granite

---

I See the next chapter for an explanation of lighthouse lenses and light sources.
II Bishop Rock has been recognized by Guinness as the world's smallest island with a building on it.

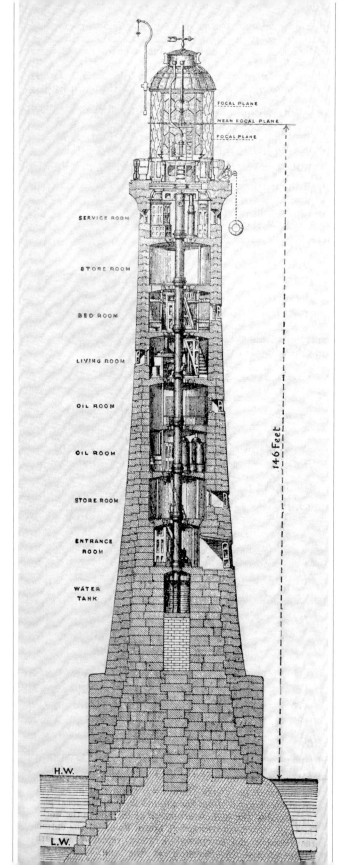

blocks was discovered, the decision was made to build a third tower better able to withstand the forces of nature. Sir James N. Douglass, who at the time was completing his masterful rebuilding of the Eddystone lighthouse, was chosen to lead the effort.

Rather than build a completely new lighthouse, Douglass elected to reinforce the 1858 tower, drawing on his recent experience at Eddystone. The base would be extensively strengthened, and the focal plane of the light elevated from 110 to 146 feet above the high-water mark. The new stonework would incorporate interlocking dovetail joints similar to those at Eddystone, hoping to avoid damage to the granite blocks that weakened the earlier tower. This beacon, constructed at what was perhaps the pinnacle of traditional nineteenth-century lighthouse technology, serves as an excellent example of the features that might be found worldwide in similar structures of the day.

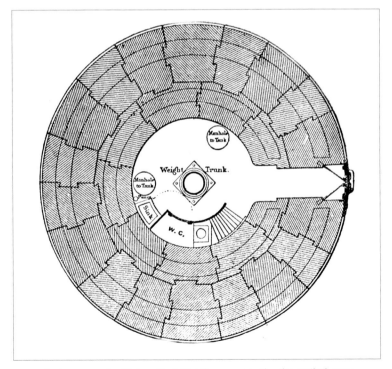

Fig. 3C Facing page, the Bishop Rock Lighthouse: note the dovetailed stone cladding surrounding the 1858 tower. Above **(Fig. 3D)**, a cross section of the Eddystone Lighthouse: note the interlocking granite stones used to build the tower and the central weight trunk that contained the weights used to power the clockwork mechanism that rotated the beacon.

First, to combat the force of the waves, Douglass built a cylindrical stone base forty feet in diameter around the foundations of the tower and extending to twenty-five feet above the high-water level. This structure and the stone cladding of the old tower above it were made of dovetailed granite. There were a total of ten levels, beginning with a water tank on the bottom and including fuel storerooms for the light, living quarters for the keepers, and a workroom and storage space near the top. Access was via an internal circular stairway. At the apex, Douglass installed large two-level lights illuminated by oil-flame lamps and projected through Fresnel lenses. The beacon rotated, turned by a clockwork mechanism that relied on power from a system of weights that were manually wound up and descended in a central trunk line. A typical cross section of similar construction from the Eddystone light provides a good illustration of these concepts. (See Figure 3D.)

The Bishop Rock lighthouse remains in service today. Oil lamps continued to be used as the primary form of illumination until they were replaced by electric lights in 1973. The tower was fully automated in 1992 and is remotely monitored. Because of its difficult location, a helipad was installed atop the light tower in 1976.

At least as fascinating, and certainly far more important than the iconic structure of lighthouses, is their function as beacons of guidance for mariners. Like construction technology, the light sources that power these beacons have evolved over the centuries, and in ways more radical than the towers that support them. These topics will be addressed in the following chapter.

CHAPTER 4

THROWING THE LIGHT

From the earliest days through the late nineteenth century, ships at sea were guided by instruments little changed from those of seafarers hundreds of years earlier. Charts, frequently outdated or simply wrong, were often of little help. While the compass, sextant, and chronometer might provide some indication of one's approximate location far from land, the rocky cliffs, hidden shoals, and shifting sands of the world's seacoasts posed a mortal danger to those brave enough to embark on a life at sea. The welcome beam of a lighthouse could make the difference between the security of a harbor and an anonymous watery grave.

The primary purpose of a lighthouse is that of a navigational aid—a distinctive landmark during daylight hours and a guiding light at night. While simple in concept, the practicalities of lighthouse science and engineering were far more complex. Problems relating to lighthouse construction were mastered long before the physics of light were understood and applied to the beacons affixed to their summits. Of equal importance were factors that could not be controlled, including weather conditions and the curvature of the earth.

The greatest progress in lighthouse technology came in the late eighteenth and nineteenth centuries as steady advances brought brighter and more efficient light sources, as well as dramatically more effective ways to project those lights over a greater distance. The lamps of Argand and the lenses of Fresnel would change the world of coastal navigation, to be supplanted only by the advent of long-distance navigational systems in the twentieth century.

The most obvious limitation to a lighthouse's effectiveness is the curvature of the earth. A light placed five feet above sea level is maximally visible only about two and half nautical miles away, well within the range of dangerous offshore shoals in many areas. Early navigational lights were located on towers, or alternatively placed on high ground to extend the range of their visibility. Elevating the light source to 100 feet above sea level extends the range to about 11.7 nautical miles, and to 150 feet to a distance of 14.3 nautical miles. But because of the earth's curvature, there is a significant diminishing return in the range of visibility as light-source elevation increases. A lighthouse 600 feet tall, for example, quadruple the height of the 150-foot tower, would have a theoretic visibility of only 28.7 nautical miles, only about twice the distance.

Rain, snow, and fog alter the transparency of the atmosphere both day and night. The visual acuity of the individual observer may play a role, as may glare from background lighting during hours of darkness. Despite these obstacles, the welcome beam of a lighthouse has guided many a ship to safety in times of distress. While the concept of achieving this goal might seem simple, the science and art of projecting light over great distances was one of the greatest problems faced by inventors, scientists, and lighthouse builders.

THE ILLUMINATION OF EARLY LIGHTHOUSES

The earliest forms of lighthouses were often little more than fires built upon seaside promontories or atop towers commonly made of stone. The illumination was provided by wood fires or, in some areas, coal or other flammable material. In many lighthouse towers, the fire was contained in a brazier, sometimes hoist up to increase elevation and visibility. This crude method had many limitations. In addition to the necessity of constantly adding fuel to the fire, the smoke generated often acted to obscure the light while seaside wind and rain could easily dampen the flames. Enclosed cupulas on lighthouses became an option as glass windowpanes came into common use in Europe in the seventeenth century. Winstanley's 1698 Eddystone Rocks lighthouse, for example, mentioned in the previous chapter, featured a glass-enclosed lightroom on its summit. Illumination was provided by a chandelier whose sixty candles needed almost constant attention.

While candles were used in some early lighthouses, a more common form of illumination was the oil lamp. Technologically simple, an oil lamp consists of a reservoir for a flammable oil, a wick dipped into the reservoir, and a heat-resistant vessel to hold both. The flame can be fed by a wide variety of fuels, ranging from vegetable oil to oils derived from animal sources. For many years, whale oil was the primary fuel for lamps used in American lighthouses. Multiple oil lamps were often placed together in "spider lamps" designed to increase the luminosity of the light source.

Fig. 4A A medieval lighthouse.

While candles and oil lamps were better suited for lighthouses than braziers or open fires, they were far from ideal. Their use of fuel was relatively inefficient, emitting unburned solids in the form of soot that, in turn, necessitated frequent cleaning of their glass-faced enclosures. Oil lamps required frequent trimming and advancement of their wicks. The simple task of hauling heavy containers of oil up a narrow stairwell to the top of a tower required substantial physical effort. But improvements were soon to arrive. The eighteenth century is often referred to as the Age of Enlightenment. Despite the lofty name, the era's advances in science and technology would extend even to the mundane necessity of household (and lighthouse) illumination in the form of an invention by a Swiss physicist named Ami Argand.

THE LAMPS OF AMI ARGAND AND THEIR SUCCESSORS

François Pierre Ami Argand was born in Geneva in 1750. Having an interest in science, he moved to Paris in his twenties to pursue a career in teaching and research. The inefficiency of the widely used oil lamp was well known, and around 1780, Argand set out to construct a better model. By 1782 he had crafted a totally new form of oil light. It still used the same readily available fuel but was significantly brighter and far more efficient.

Argand's invention addressed several important problems with existing oil lamps. Rather than using a single woven wick that drew oil to the flame by capillary action, his wick was circular and enclosed between two metal sleeves around a hollow center. A good flow of air, necessary for the complete combustion of the oil, was thus achieved. Fuel was delivered by gravity via a tube from a separate reservoir located above the level of the wick. Turning a knob on the side of the lamp advanced the wick itself. To further increase airflow and protect the flame from wind, a glass chimney was added. The improvement was dramatic. The new lamp was said to be as bright as six to eight candles while using less fuel. The increased airflow provided more oxygen to the flame, significantly reducing soot.

While far better than preexisting oil lamps, Argand's invention had some drawbacks. The location of the oil reservoir partially blocked the lamp's light. As the oil was gravity-fed, variations in oil density had the potential to slow fuel supply and limit the types of oil that could be used. Argand's wick was simple to advance but still required manual adjustment. Over the next century, other inventors crafted modifications and improvements to his design, allowing Argand's core

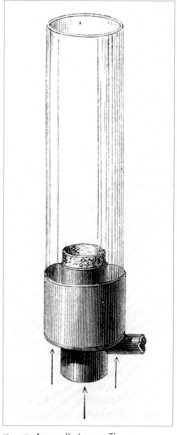

Fig. 4B Argand's Lamp. The arrows indicate the direction of airflow. The tube to supply oil to the wick is seen at the lower right.

concept to remain the major source of lighthouse illumination well into the late nineteenth century.

INTENSIFYING THE LIGHT

While Argand's lamp significantly increased the brightness of artificial light, a fundamental problem remained unsolved, that of projecting this light over great distances. Light emitted from a single source such as a candle or oil lamp radiates in all directions. If the goal is to project the light out to sea, only that portion of the light directly visible from the sea is of value. In a sense, the rest of the light emitted from the lamp is wasted. An obvious solution seemed readily apparent: place a reflective surface behind the lamp to force the light's rays in the desired direction. This solution had been attempted at least as early as the sixteenth century, when polished metal reflectors were placed behind the open fires of early lighthouses.

By the eighteenth century, the basic physics of light reflection were understood, but the technology of crafting efficient mirrors was not yet available. Glass mirrors, although initially small in size, had been in common use since the time of the Roman Empire. All mirrors absorb part of the light that strikes their surface while reflecting the remainder at angles that correlate with the position of the light source in relation to this surface. Briefly stated, a flat mirror (or other reflective surface) scatters light without directing it in one specific beam. A spherical mirror (i.e., a mirror made from a section of a round globe) reflects light back at its source. A parabolic mirror, when properly aligned with a lamp or other light source, is able to reflect much of the light in a single direction, effectively increasing the directional brightness of the light. A graphic representation of this concept can be seen in Figure 4C.

The earliest parabolic reflectors were made of handcrafted metal or small pieces of flat mirrored glass affixed to a backing that was as close to parabolic as the technology of the day would allow. As the flame needed to be at the focal point of the reflective surface, one modification of Argand's lamp placed the oil reservoir behind the reflector. (See Figure 4D.) Together, the system formed by these

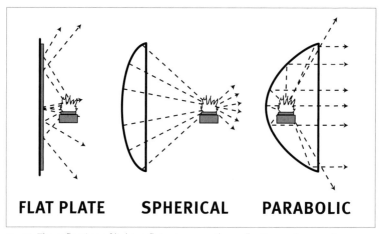

Fig. 4C The reflection of light: a flat mirror or other reflective surface scatters most light. A parabolic reflective surface is capable of channeling the light that strikes it into a more focused beam.

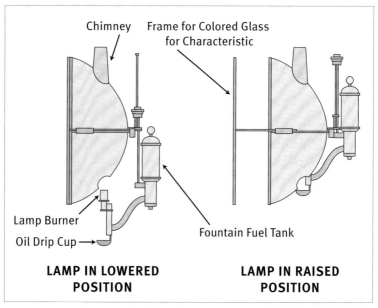

Fig. 4D Stevenson's retractable light reflector: Stevenson, of the famed family of Scottish lighthouse builders, designed this variation on the Argand lamp and parabolic reflector in about 1810. The lamp could be lowered for cleaning and adjustment, while the fuel reservoir was behind the reflector so as not to block any of the emitted light. In order to add identification to the light, a piece of colored glass could be placed in front of the unit.

advances remained the standard for lighthouse illumination until the invention and general use of Augustin Fresnel's lens.[I]

While reflectors served to provide a small increase in the amount of light beamed in a given direction, an ingenious English glassmaker named Thomas Rogers conceived of attempting to concentrate the light through the use of a lens. Perhaps inspired by the so-called "magic lantern" that had been used since the 1600s to project images on a flat surface, Rogers reasoned that by placing a planoconvex lens *in front of* a lamp, the light's beam could be focused into the distance, thus increasing its range.[II] Rogers's lenses were used in several English lighthouses in the late eighteenth century. While marginally improving the power of the light emitted, much of the potential value of the lenses was lost due to the technical difficulty in forming a precise shape and the thickness of the lens, which absorbed a significant amount of the transmitted light.

WINSLOW LEWIS'S LAMP

Any discussion of lighthouse illumination in the new republic of the United States must include Winslow Lewis. He was born in 1770 into a seafaring family from Massachusetts. The young Winslow followed the sea as well, serving as a ship's captain for more than a decade plying the lucrative transatlantic route between Boston and Liverpool. The Embargo Act of 1807, an ineffectual attempt to force warring European nations to respect American neutrality, brought an abrupt halt to Lewis's chosen career. Knowing the sea well and having a streak of entrepreneurship, Lewis believed he could invent a more effective navigational light source.

After two years of work, Lewis applied for a patent on what he termed "a new and improved Magnifying and Reflecting Lantern

---

[I] Light emitted via reflection is termed *catoptric*, whereas light whose path has been altered by passage through a lens (i.e., refracted light) is termed *dioptric*. Light that is both reflected and refracted is termed *catadioptric*.

[II] A "magic lantern" was an early form of a slide projector, reputedly invented by Christiaan Huygens, a Dutch scientist. The concept remains in common use today via the projection of digital images illuminated by LED-generated light. Magic-lantern shows were popular in America prior to advent of motion pictures.

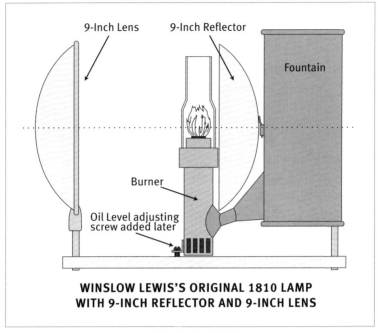

Fig. 4E Winslow Lewis's Lamp

for use in lighthouses." Designed to burn "sperm oil,"[III] the lamp featured a circular wick, a circular glass chimney to increase airflow to the flaming wick, "a reflector which gathers the light together," and a "Bull's Eye" planoconvex lens in front of the flame designed to focus the light and extend its range. In his patent application, Lewis noted the lamp would consume "two gills and a half in twelve hours if the Spermeciti Oyl [sic] be good." The new light was designed to be mounted on racks and used in multiple arrays. He was granted Patent No. 1305 on June 8, 1810.

If Lewis's lamp seemed similar to that of Argand, it was. Lewis claimed having no knowledge of Argand's invention, despite having spent a number of years sailing back and forth to Europe, where such lamps were in common use. He claimed that he devised the design completely on his own. As one author noted, "Though there is no clear proof as to the true origins of Lewis's 'invention,'

[III] Sperm oil: A type of "whale oil" derived from sperm whales. See glossary for a more complete explanation of these terms.

it strains credulity to think that Lewis, a man who admittedly had no optical experience whatsoever, had come up with this design on his own. It is virtually certain that he copied something that he had seen. But whether he did it or not, one thing is absolutely true—Lewis's lighting apparatus was not very good."[1]

Lewis's version of Argand's lamp was of relatively poor quality. The reflector that he claimed to be parabolic was not and, moreover, was said to have low reflectivity. The lens that Lewis installed in front of his lamp to focus the light was, like the lamp, not his original idea. It was essentially the same as that of Thomas Rogers and suffered from the same limitations in its transmission of light.

Despite these negatives, Lewis's light was in many ways superior to most other lighting systems in use in America at that time. Armed with a patent and a good sense of politics, Lewis became a major name in American lighthouse illumination and construction over the following decades. Light designs based on Lewis's patent were used in most American lighthouses until the 1850s. More details on this will follow in the chapters on American lighthouses.

THE MASTER OF LIGHT, AUGUSTIN FRESNEL

Arguably the greatest advance in lighthouse technology came from the work of Augustin-Jean Fresnel, a French physicist and engineer.[IV] Fresnel was born in Normandy, the son of an architect. He was said to be a sickly child but was recognized at an early age for his intelligence and curiosity. At age sixteen, he enrolled in France's prestigious École Polytechnique, continuing his education after graduation at the École Nationale des Ponts et Chaussées, where he trained as a civil engineer. Assigned to the often-mundane task of building roads and bridges in rural France, Fresnel's attention turned to the study of the nature of light. It had long been observed that light can be both reflected from shiny opaque surfaces and refracted by passing through translucent substances.[V] The leading—and competing—theories of the day

---

[IV] Fresnel is pronounced *frey-NEL*.

[V] Refraction, as applied to visible light, refers to the change in direction of light waves when they pass through a translucent medium. The most common example is eyeglasses, which more accurately refocus light on the retina of the eye to improve the sharpness of a visual image.

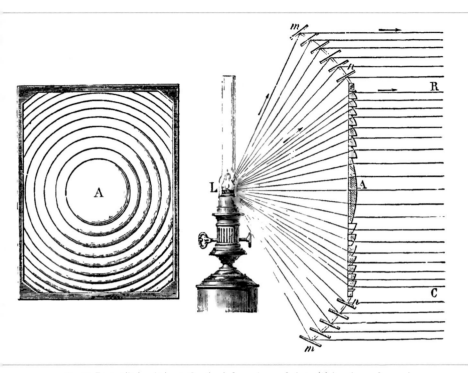

Fig. 4F Fresnel's basic lens: On the left, a piece of glass (A) has been formed such that a central bullseye is surrounded by a concentric series of stepped prisms. On the right, the light from the lamp (L) strikes rear of the lens (seen here in transverse section) and is focused in a single direction. To increase efficiency, angled mirrors (M) at the top and bottom of the central lens reflect the light in the same direction as the light that emerges from the lens. As the projected light is both refracted by the lens and reflected by the mirrors, this would be termed a catadioptric light source.

were those of Dutch physicist and astronomer Christiaan Huygens and English scientist Sir Isaac Newton. Huygens believed that the known properties of light were best explained by considering it as a series of "undulations," or waves, a theory based on mathematics. Newton, on the other hand, felt the observed physical properties of light led to its being described best as a form of weightless particles, which he termed "corpuscles." The scientific controversy of the day revolved around the search for a coherent explanation for diffraction, the observation that a beam of light passed through a narrow slit or past the edge of a solid object can produce a pattern of light and dark lines. Fresnel set out do his own observations and develop his own theories.

Through his interest in the physics of light, Fresnel had become acquainted with some of France's leading scientists. In 1817, the Académie des Sciences announced that its annual prize would be awarded for the best paper submitted on the problem of diffraction. Fresnel's paper presented a detailed mathematical explanation for describing and predicting patterns produced by diffracted light based on Huygens's wave theory. Despite opposition from such leading scientists as Pierre-Simon Laplace, the Académie awarded Fresnel the Grand Prix.

Now recognized as an expert in the science of light, Fresnel was recruited to advise the national Commission des Phares on improving lighthouse visibility. French lighthouses of the day were illuminated by Argand lamps with various types of reflectors, sometimes augmented by glass lenses of the type Thomas Rogers had used in England. Attempts to focus and magnify light through these lenses were only marginally effective because the necessarily thick glass absorbed a significant amount of the emitted light. Based on his understanding of how the direction of light waves can be altered by passing through translucent objects, Fresnel reasoned that a series of thin lenses arranged in a bullseye pattern would be able to focus the light in a single beam while minimizing light absorption by the lens itself. This rudimentary idea was soon augmented by adding reflective mirrors at the periphery of the lens to capture and direct an even greater amount of light.[VI] (See Figure 4F.)

To bring his idea to reality, one of Fresnel's major challenges was finding someone capable of making the high-quality glass needed while willing to invest both the time and funds in a potentially risky endeavor. With the assistance of François Soleil, a small Parisian precision-instrument maker, over the next few months Fresnel built his first prototype lens, a square glass panel measuring roughly fifty-five centimeters on each side and composed of an inlaid pattern of individual glass prisms in a bullseye configuration. Members of

VI While Fresnel is appropriately given credit for the lens that bears his name, it should be noted that the concept of a flat stepped lens had been conceived by at least three prior individuals, most notably Georges-Lewis Leclerc, the Comte de Buffon, in 1748. Buffon, a French polymath, suggested that such a lens would increase light transmission by reducing the thickness of the glass. Neither Buffon's nor either of the other earlier concepts found a practical use until Fresnel independently suggested such a lens would be ideal to focus a lighthouse beam.

the Commission des Phares were most impressed. They requested that he make a full working prototype similar to one that would serve as a lighthouse's beacon.

While building his lens prototype, Fresnel soon realized the need for a lamp that emitted more light than the standard Argand lamp and reflector. Working with another physicist, François Arago, the two devised a variant on Argand's lamp. Instead of a single circular wick, the Fresnel-Arago lamp had four separate concentric wicks around a central airflow shaft. The result was a lamp that was said to be twenty times brighter than any other light source available at the time.

Fig. 4G Augustin Fresnel

By the spring of 1821, Fresnel's mockup of a lighthouse light source was ready to be demonstrated. It was octagonal in shape, with eight separate bullseye lenses surrounding the new, brighter lamp. On April 13th, Fresnel's apparatus and two other conventional light sources were set up at the Observatoire de Paris, just south of the Luxemburg Gardens. The observers were nearly four miles to the north on the hill of Montmartre. The entire Commission des Phares as well as a number of sailors were present to judge the new light. The dramatic superiority of Fresnel's invention was obvious to all. The next chapter in lighthouse illumination had begun.

With a new and more powerful light source available, the commission conceived a bold plan to establish a series of lighthouses along the entire French coast so that mariners would never be out of sight of a known light. Fresnel's lamps and lenses, of course, would generate all the new beacons. It was an ambitious plan, and Fresnel dove into it eagerly. As not all locations would require the same type of light, Fresnel produced six sizes of lenses, divided into orders based on their size and focal length.[VII] These

---

[VII] In order to most efficiently capture the light from a lamp or other source, the source needed to be exactly positioned at a certain distance from the lens, i.e., the focal point. This distance varied based on the lens size and configuration.

ranged in size from a height of nearly eight and one-half feet for a first-order lens to only about a foot and a half for the smallest, a sixth-order lens. To avoid confusion between lighthouses, Fresnel proposed that some lights be constant while others rotate or flash in a known pattern. To achieve this goal, he worked with a clockmaker to devise a mechanism to rotate his lens apparatus around a fixed light source.

The first of Fresnel's lights was scheduled to be installed atop the Cordouan light, located on a small island at the mouth of the Gironde estuary in the Bordeaux region. It was France's oldest lighthouse, put into operation in 1611 and originally illuminated by a wood-fueled fire on its summit. The light had been modernized but still was not ideal in its role of guiding sailors through this dangerous passage. The new light was to be tested in Paris from atop the as-yet-unfinished Arc de Triomphe in the summer of 1822. Fully living up to its expectations, observers could see it clearly from a village twenty miles north of Paris. With the commission's approval, the light was disassembled and moved to Bordeaux, where it first illuminated the Cordouan tower in July 1823.

In the midst of this, Fresnel's health began what would become a steady decline. It became clear that he suffered from what was then known as consumption, an all-too-common wasting disease of the era that today is recognized as pulmonary tuberculosis. Now acclaimed nationally for his genius, he continued his scientific work on the nature of light. In 1823, Fresnel was elected to the prestigious Académie des Sciences and the same year received France's highest award, the Legion of Honor.

As Fresnel's illness worsened, his younger brother, Léonor, began working with him to complete the task of illuminating France's coast. It was a role that he easily adopted. Léonor's background and education were quite similar to those of his brother. By 1826, Augustin's health had reached the point that he was no longer able to do productive work. He died at age thirty-nine in July 1827. Over the next two decades, Léonor would continue and expand his brother's legacy, instituting variations and improvements on lighthouse illumination systems that would remain the world's standard well into the twenty-first century.

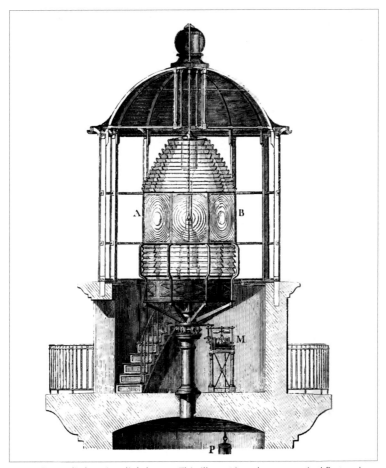

Fig. 4H Fresnel's lens in a lighthouse. This illustration shows a typical first-order lens designed to cast a rotating beacon. The central bullseye lenses (A and B) are situated between banks of prisms at the top and bottom of the light. The entire light mechanism is mounted on a rotating platform powered by a wind-up clocklike mechanism (M), which is driven by a weight (P) that descends through a shaft in the lighthouse tower. Alternative configurations designed to emit a different light pattern featured an outer frame that could rotate around a fixed light source.

NEW FUELS AND NEW WAYS TO GENERATE LIGHT

Although variants of Fresnel's original lens design remained the standard for lighthouse illumination, the sources of light they directed changed with time and technology. In 1819, it was estimated that of the 254 lighthouses in use in Europe, "5 were still lit by candles, 30 by wood or coal fires, 2 with coal gas, 157 with common oil lamps, and 60 with Argand lamps."[2] Through the middle of the nineteenth century in North America, sperm oil was the preferred fuel. With time, however, it had become increasingly expensive due to an expanded demand for its other uses and a dwindling supply as unrestrained whaling decimated whale populations. Lighthouse authorities sought and often switched to other fuels including vegetable oils and oils rendered from animal fat. The discovery of petroleum near Titusville, Pennsylvania, in 1859 offered some respite to the beleaguered whales while raising the possibility of inexpensive kerosene (then referred to as "mineral oil") as fuel for lighthouse lamps. Gas, most commonly generated from coal, found limited use, primarily because of the necessity of building an on-site generating plant to manufacture it.

Acetylene, whose use as a light source was primarily due to the work of Gustaf Dalén, a Swedish industrialist, provided a brilliant white light. Importantly, Dalén devised a way to automatically turn the light on and off, making it well suited for remote lighthouses and unattended buoys. Ironically, he was blinded by an acetylene explosion in 1912 but later that year was awarded the Nobel Prize in Physics for his "invention of automatic regulators for use in conjunction with gas accumulators for illuminating lighthouses and buoys."

The promise of a relatively unlimited source of electric power became a reality in 1831 with Michael Faraday's invention of the magneto-electric generator. By the latter third of the century, electricity was emerging as the power source of the future and an obvious replacement for other types of lighthouse illumination. The major obstacles to its use were the often-remote location of the lighthouses and the necessity of a separate power source (such as steam) to turn an electrical generator. In the 1880s, Trinity House, the official lighthouse authority for Great Britain, conducted a study to assess the relative superiority of oil, gas, and electricity as illuminants. Based on more than 6,000 observations, the results clearly indicated electricity as the power source of choice, though—at that time—oil was found to be more economical. The arc light, first used in an English lighthouse in 1858, provided a brilliant light but was difficult to maintain. Incandescent lightbulbs would eventually become the standard for the following decades, coming into

Fig. 41 The use of whale oil as lighthouse fuel (among other applications) seriously depleted whale populations worldwide. The caption of this cartoon from the December 31, 1860, issue of "Vanity Fair" reads, "Grand Ball given by the Whales in honor of the discovery of the Oil Wells in Pennsylvania."

general use in lighthouses around 1920 and thereafter. One of the most famous, and perhaps unusual, early lighthouses to be illuminated by electricity was Auguste Bartholdi's "Liberty Enlightening the World," better known in the United States as the Statue of Liberty at the entrance to New York Harbor. Bartholdi's creation was initially promoted as having a dual purpose, described as "a combined goddess and lighthouse," by one writer.[3] The sculptor's original vision was inspired by a visit to the Middle East in the 1850s, where he became fascinated with large-scale public monuments such as the Sphinx and Pyramids of Giza in Egypt. In the 1860s, the government of Egypt expressed an interest in a lighthouse at Port Said near the Mediterranean entrance to the Suez Canal. Eagerly, Bartholdi designed a "colossal statue of a robed woman holding a torch," to be named "Egypt Carrying the Light to Asia,"[VIII] but the Egyptians rejected the concept on the basis of

VIII An alternative name was "Progress Carrying the Light to Asia."

cost. A far more conventional lighthouse designed by French architect François Coignet was built instead.

In 1871, Bartholdi visited the United States with the seed of an idea to construct a similar colossal statue near New York Harbor. Over the next decade, support was garnered for the project, and funds were raised in both the United States and France. In 1877, Congress passed a joint resolution in support of the project, now known as the Statue of Liberty. On October 28, 1886, more than a decade later, the statue, formed of copper sheets crafted by Bartholdi over a rigid metal framework built by the firm of Gustave Eiffel, was dedicated on Bedloe's Island, its new home.

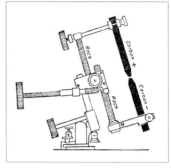

Fig. 4J A typical art light, circa 1900. Arc lights generated light by the arcing of electric current between two carbon rods.

From the very outset, the lighting configuration of the new structure presented a series of problems. How should the upraised torch and statue itself—made of dark and poorly reflective material—be illuminated? If the statue was to serve its secondary purpose as a lighthouse, what was its visual range? Was it of navigational assistance to ships entering the harbor? There was universal agreement that the lights should be powered by the relatively new technology of electricity. This involved building a steam plant to power the generator, which, in turn, produced the electricity. On November 16, 1886, after a number of institutional problems and false starts in lighting the torch and statue, President Grover Cleveland sent a letter to the secretary of the Treasury directing "that the statue of Liberty Enlightening the World be at once placed under the care of the Light-House Establishment, and that henceforth it be maintained by the Light-House Board as a beacon, under the regulations pertaining to such beacons."[IX] Two days later, the board duly issued a "Notice to Mariners" announcing the transition.

[IX] The interested reader will note that occasionally the word for these beacons is written as "light-house" (or other variant), usually within quotation marks. For the sake of clarity, however, I have generally chosen to use the current spelling of "lighthouse" in other instances even when the original term was hyphenated.

The change was a positive one in that it moved the lighting problem into the realm of those with knowledge and experience in that field. The initial illumination of the statue was provided by arc lamps while efforts continued to give it value as a navigational aid. Surveys of seafarers revealed that under ideal conditions, the light was visible as far out as twenty-three miles at sea, but the torch was not especially useful in determining position. Despite various modifications to the lighting system, by the early 1890s it was clear that the colossal statue was of little practical use for navigation. A report submitted in early 1894 by the Lighthouse Board's Committee on Lighting concluded that the Statue of Liberty "is of no importance whatever" as an aid to navigation. In 1902, supervision and maintenance of the monument was turned over the War Department, and in 1933 to the National Park Service, under whose care it remains today.

By the first quarter of the twentieth century, most lighthouses had been converted to electrical illumination. Although still having importance as navigational aids, their usefulness as such was at first augmented by, and later supplanted by, radio and other forms of wireless directional markers.

WHICH LIGHTHOUSE?

While the purpose of lighthouses was that of guides and warning beacons for mariners, history abounds with stories of ships lost because their captains mistook one lighthouse for another. This was not a major problem in America until well into the nineteenth century; there were simply too few lighthouses to confuse one with another. Fresnel recognized this potential danger as the Commission des Phares pursued its plan to vastly increase the number of French coastal beacons. He developed, as did others, ways to make a lighthouse's beacon unique. His use of rotating lights and lights that flashed in a distinct pattern would become vital in identifying individual lighthouses.

During the 1850s in the United States, the Lighthouse Board began painting lighthouses with distinctive patterns to make them more readily identifiable during periods of daylight. The bold three-band, black-and-white pattern of Tybee Island's light and the distinctive red-and-white horizontal stripes of Sapelo Island's lighthouse are good examples of this.

LATER ADVANCES IN NAVIGATIONAL TECHNOLOGY

Although certainly not recognized at the time, Guglielmo Marconi's demonstration of the practicality of wireless signal transmission to the British government in July 1896 sounded the death knell for lighthouses as practical navigational aids. It soon became apparent that the transmission of visible light—a form of electromagnetic wave—was to be supplanted by the transmission of radio signals, another and far more versatile form of the same principle of physics.

Progress came rapidly. Soon wireless telegraph communication was standard for ships at sea. In 1906, the first radio detection finder was demonstrated. In 1921, the first radio beacon installed. The 1930s saw the practical development of radar. LORAN (LOng RAnge Navigation), a radiolocation system with a range of 1,500 miles, was developed by United States military during World War II. By the 1960s a form of satellite-based location system was demonstrated, leading to the practical implementation of global positioning systems in the 1980s. Perhaps sadly, much of the progress in wireless location systems during the twentieth century resulted from military research. Today, highly accurate positioning systems capable of locating an individual's position within a radius of meters are found in the ever-ubiquitous cell phone.

The advances in wireless-based location systems have been matched by advances in the technology of light production and transmission. While gas (chiefly propane) and acetylene still find limited use in some navigational lights, most modern lights rely on high-output incandescent, metal halide, or light emitting diode (LED) sources. Not only are these lights more efficient in producing brighter illumination at a lesser energy cost, they are generally far smaller and have a much longer useful life than the sources they replaced. Advances in energy storage, especially lithium-based batteries, have allowed many lights to rely wholly or in part on solar- and wind-based electrical generation.[4]

Fresnel's lenses remain the standard for navigational lights. In sharp contrast to their original high cost and laborious production process, however, they are now formed almost exclusively from relatively inexpensive molded plastic. This advance, coupled with the availability of smaller and more efficient light sources, has allowed a dramatic reduction in the size of navigational lights. For example, a typical rotating first-order Fresnel lens apparatus dating from about 1900 measures approximately 4.5 meters (just under 15 feet) in height. An equivalent modern beacon with six symmetrical Fresnel lens panels measures 0.7 meters (about 2.3 feet) in height.[5]

THE PAST AND THE FUTURE

It is reasonable to say that the era of lighthouses has long passed, made obsolete by the relentless march of science and technology. Yet these majestic sentinels remain to remind us of man's past and continuing quest to understand and conquer the challenges that nature has laid before us. They represent a small but important symbol of our quest to explore the far reaches of our own world, just as evolving technology continues to aid us in our attempt to explore the far heavens above our earth.

CHAPTER 5

AMERICAN LIGHTHOUSES

"The United States built a lighthouse system that today ranks with the best in the world, but it was not an easy growth, and ineptness, conservatism, poor management, penury, political patronage, technological backwardness, and waste marked its path before the country began steering the course that led to a great lighthouse system."

Francis Ross Holland Jr., *American Lighthouses* (1972)

COLONIAL LIGHTHOUSES

From the earliest days, Britain's North American colonies were a group of semi-independent settlements established for the purpose of expanding her role in the New World and beholden to an oft-disinterested monarch thousands of miles to the east beyond the turbulent North Atlantic Ocean. As transoceanic commerce expanded in the early eighteenth century, the perils of the sea, as marked by shipwrecks, with their attendant loss of life and goods, became a major issue in the colonies, as it did in the home country during the same period. The ports of America's Eastern Seaboard from Virginia to Massachusetts—the main centers of population in those early days—became hubs of trade and wealth for many merchants, shipowners, and investors. And as in Britain, these losses represented a major and potentially avoidable risk that might in part be alleviated by improved navigational support in the form of lighthouses. The colonies were expected to be self-supporting. The chance of Parliament voting appropriations for such was remote at best. Consequently, the impetus for the construction of America's first lighthouses often came not from governmental concern but rather from those with a pecuniary interest in the shipping trade.

America's first lighthouse was built on the treacherous approach to Boston Harbor. In 1715, in response to a petition from local merchants, the General Court authorized the appropriation of funds to build a lighthouse on Little Brewster Island. Construction costs and ongoing maintenance were to be paid by tolls levied on passing ships. A keeper was hired, and the light put into operation in September 1716. The light functioned well, surviving a fire in 1751 and several lightning strikes over the years, all repaired in good order.[1] Though like many lighthouses in time of war, the original tower met its end during the American Revolution.

As the Little Brewster light served as a guide to Boston Harbor, in 1775 American troops removed the lamps and set the tower on fire. The British quickly took control of the island and made repairs, only to have a contingent of American troops promptly retake it. The British regained control of the lighthouse during their occupation of Boston, but on their retreat in June 1776 blew it up to deny the American side its use. It was rebuilt in 1783, and in 1859 elevated by fourteen feet to stand 102 feet above sea level. In the twentieth century, as lighthouses were being automated and resident keepers were being phased out, the Little Brewster light was scheduled to be the last in the United States to be decommissioned. As a historical structure, however, it remains under the administration of the Coast Guard as the last manned lighthouse in the United States. It was fully automated in 1998.

Among the dozen American lighthouses in existence at the time of the Revolution, their navigational value made them strategic targets for both sides. The 1764 Sandy Hook Light, built in New Jersey on the approach to New York Harbor, became a target of American forces during the conflict. An effort was made to destroy it, but the tower was so well constructed it remained intact, making

---

[1] In his *Lighthouses and Lightships of the United States* (1917), George R. Putnam quotes a 1789 writer who noted that after the lighthouse was struck by lightning several times, "attempts were made to erect conductors; but this measure was opposed by several of the godly men of those days, who thought it vanity and irreligion for the arm of flesh to presume to avert the stroke of Heaven." They eventually gave in and allowed lightning rods to be installed.

it today the nation's oldest surviving lighthouse. Cape Henlopen Light, constructed in 1767 in Delaware on the approach to the port of Philadelphia, was burned by British troops and out of operation until after the war. Gurnet Light,[II] built in 1769 on the entrance to Plymouth Bay in Massachusetts, was hit and damaged by a cannonball in a 1778 artillery exchange between a British warship and American militia. The 1771 Cape Ann Light on Thatcher Island near Rockport, Massachusetts, is significant in that it was the last light station to be established under colonial rule and the first American lighthouse to mark a navigational hazard rather than a harbor entrance.[1] An important guide for shipping, its keeper was a well-known Tory opposed to the cause of American independence. As a result, local patriots deposed him from his job, thus leaving this vital light dark for the remainder of the war.

A NEW AMERICAN ERA BEGINS

At the end of the Revolutionary War, there were a total of twelve Colonial-era lighthouses, as well as four more under construction by the time the new American Constitution was ratified and the federal government assumed power in 1789. When the first new Congress met during the spring and summer of that same year, it passed, in a joint session of the House and Senate, *An Act for the establishment and support of Lighthouses, Beacons, Buoys and Public Piers*. It was the ninth legislative action of the new republic, an indication of its relative importance. The Act specified that the central government would be responsible for the expenses and upkeep of these navigational aids if the former colonies were willing to cede ownership. The newly federally owned towers would be administered under the Treasury Department. As an indication of the miniscule size of the government in those early days of the nascent nation, contracts were to be "approved by the President of the United States," who also had responsibility for appointing personnel to operate and manage the lighthouses. The states were in no hurry to cede control of such valuable assets; it would be 1797

II Also known as Plymouth Light.

before all of the existing lighthouses came under federal control. Over the following years, their number grew rapidly. By 1800 there were twenty-four, all on the Atlantic coast; by 1820 the number would increase to more than fifty.

During the first years of the republic, the Lighthouse Establishment, as it came to be called, was under the direct attention of Alexander Hamilton, secretary of the Treasury.[III] The records of the National Archives include correspondence between Hamilton and President George Washington regarding relatively minor details of day-to-day administrative issues. As the number of lighthouses increased and the burden of management grew, Hamilton turned over supervision of lighthouses and related navigational aids to the commissioner of Revenue within the Treasury Department. Over the next few years, oversight bounced back and forth between the commissioner and the secretary of the Treasury. In March 1812, in an apparent effort to bring some uniformity and simplify maintenance and repairs, Congress authorized the secretary of the Treasury,

> to purchase of Winslow Lewis, his patent right to the plan of lighting lighthouses, by reflecting and magnifying lanterns, if the same shall be proved to be a discovery made by him; and to contract with the said Winslow Lewis, for fitting up and keeping in repair, any or all the lighthouses in the United States or the territories thereof, upon the new and improved plan of the reflecting and magnifying lanterns; or to contract with the said Winslow Lewis, for such sum as he may think for the interest of the United States.[2]

In essence, Congress had outsourced the lighting and maintenance of the nation's lighthouses to a former sea captain with no formal training in either optics or engineering. All lighthouses in the United States were to be retrofitted with lamps made under Lewis's patent, considered by many to be of inferior quality. As Lewis's close involvement with the Lighthouse Establishment grew, he bid on and was awarded contracts for the construction of many of the country's new lighthouses, as well as supplying oil

---

[III] More formally, the United States Lighthouse Establishment (USLHE).

to burn in his lamps.^{IV} These decisions would have far-reaching implications.

In 1820, a second management decision was made that would influence American lighthouse quality for more than three decades. When the Office of Commissioner of Revenue (which was in charge of the lighthouse service at the time) was abolished effective July 1, 1820, management responsibility was turned over to the Fifth Auditor of the Treasury, a man named Stephen Pleasonton.^V The details of the early life of Pleasonton are unknown. He was believed to have been born in 1776 and appears to have worked as a government employee during his entire career. He was known to be a clerk in the State Department when the United States government moved from Philadelphia to the new capital of Washington in the spring of 1800. Other than managing the Lighthouse Establishment for some three and half decades, Pleasonton's greatest claim to fame happened in 1814 when he saved the original copy of the Declaration of Independence from destruction by British troops as they burned much of the capital. In anticipation of the British occupation of Washington, James Monroe, at the time secretary of state, tasked Pleasonton with gathering and preserving the books and papers of the State Department, including the Articles of Confederation, the Constitution of 1789, treaties, and other irreplaceable documents dating to the foundation of the republic. As Pleasonton was about to flee the city with the documents, he noted that the Declaration of Independence had been overlooked and was still hanging in its frame on the wall. Taking this as well, he saw that his cargo was carried some thirty-five miles away to safety in Virginia, thus preserving the nation's treasures. When Monroe was elected and took office as president in 1817, he appointed Pleasonton to the

---

IV The lighthouse at Sapelo Island, Georgia, dating from 1820, is one of the few remaining lighthouses constructed by Winslow Lewis. Its history is examined in more detail in Chapter 10.

V The title "Fifth Auditor" may be confusing. By way of explanation, at the time, the Treasury Department had several senior accountants, each with a specific area of responsibility. They were designated as the First Auditor, the Second Auditor, etc. Financial oversight of Lighthouse Establishment was among the duties of the Fifth Auditor.

Fifth Auditor's position, perhaps as a reward for his service three years earlier. He would hold this position until his death in 1855.

Pleasonton's portrait from the Library of Congress is that of a dour man with a long nose, low-set ears, and a seemingly permanent frown. One author described him as

> zealous, hardworking, conservative, and an overly conscientious guardian of the public dollar. As Fifth Auditor he was one of the nation's principal bookkeepers, and he brought to the job of general superintendent of lights the bookkeeper's lack of imagination. He did not have a maritime background or any other experience that suited him to take over the country's aids to navigation.[3]

The nominal role of general superintendent of lighthouses was one among many of Pleasonton's duties as Fifth Auditor. Being based in Washington and lacking both the training and time to become closely involved in day-to-day management, on a practical basis he delegated this role to the local Collectors of Customs, who became administrators for the lighthouses in their individual districts. They were in charge of selecting the sites of new lights, the monitoring of the contractors who constructed them, arranging for repairs if needed, etc. Management of the lighthouse keepers was the collectors' responsibility as well, although the actual appointment was made through the secretary of the Treasury. As compensation for their service, they were paid a commission of 2.5 percent of the funds expended on lighthouses in their districts.[4] During Pleasonton's supervision from 1820 until the formation of the Lighthouse Board in 1852, approximately three hundred new lighthouses were built.

Under Pleasonton's oversight, critics of the management of the Lighthouse Establishment grew increasingly vocal. Word of Augustin Fresnel's new lens lighting system reached American shores in the 1820s even as current and new American lighthouses continued to be equipped with Winslow Lewis's inferior lights. Concerns were expressed over Pleasonton's reliance on Lewis for the construction of new lighthouses, many of which were later found to be shoddily built. The fact that new light locations were determined by Collectors of Customs rather than experienced seamen became a point of contention. The publishers of *The American Coast Pilot*,^{VI} an influential sailing guide for mariners, became prominent critics of the lighthouse system and its management. The concerns prompted Congress in 1837 to call for a general survey of how the nation's navigational system was being managed. This, in turn, led the following year to a formal inspection of the system by experienced naval officers charged with reporting their findings back to Congress.

Fig. 5A Stephen Pleasonton, the Treasury Department's Fifth Auditor and manager of the Lighthouse Establishment, 1820–1855.

The report of the naval officers revealed a host of problems. These included less-than-optimal placement of coastal lights, marked variations in the maintenance of both lights and lighthouses, and concerns with construction quality and the techniques used to build the lighthouses. Problems in personnel and maintenance included not only lighthouses but extended to lightships as well. It was suggested that Congress restructure the Lighthouse Establishment, but no action was taken.

About the same time, the Senate passed a resolution directing Pleasonton to assess the value of transitioning from Lewis's lamps to the superior Fresnel lighting system. Two such lights meant to serve as demonstration models were ordered from France and installed in the twin towers of the Navesink lighthouse in New Jersey. Even Pleasonton agreed that the lights were superior but decided not to purchase them, averring that their cost outweighed their benefits. As one lighthouse historian pointed out,

---

VI *The American Coast Pilot* was first published in 1796 by Edmond Blunt in Newburyport, Massachusetts. It remains in publication in the form of the *United States Coast Pilot*, published annually by the Office of Coast Survey, a part of the National Oceanic and Atmospheric Administration's National Ocean Service.

The fact of the matter was, that while the initial cost of a Fresnel lens was greater than a reflector system, in the long run it was considerably less expensive. First, a primary seacoast lighthouse might have twenty-four lamps with reflectors each burning oil, where a Fresnel lens relied on a single lamp thus resulting in a considerable savings of oil, not to mention wick material, lamp replacements and time. Secondly, the silver on the face of the copper parabolic reflectors eventually wore out from polishing, where the Fresnel lens was basically good forever. But finances aside, the lenticular Fresnel system was vastly superior to [Lewis's] reflector system and Pleasonton just couldn't see the value of importing them.[5]

Pleasonton's defense of the status quo was based on the assertion that in his primary role as a bookkeeper he had managed the system of lighthouses, lightships, and other navigational aids efficiently and at a relatively low cost to the government. There were those who criticized his reliance on Winslow Lewis, but no malfeasance was ever established.

Despite the negative report of 1838 and Pleasonton's rejection of the Fresnel lens, Congress remained unwilling to act. In 1842, the House of Representatives passed a resolution calling on the secretary of the Treasury to launch an investigation into the expenditures of the Lighthouse Authority. In response, the secretary of the Treasury appointed a Boston civil engineer, I. W. P. Lewis, to lead the inquiry. In addition to his formal educational qualifications, Lewis was the nephew of Winslow Lewis. His report, delivered to Congress in early 1843, was even more detailed and damning than the previous investigation of 1838. In the precise and detailed words of an engineer, Lewis described a broken system, administered poorly and haphazardly, in need in repair and improvement, and in desperate need of reform. A "prominent Boston journal" described the survey as "a severe blow to the defenders of the old system...; if the government had possessed the proper energy and vigilance, such an array of facts could not have been passed over unnoticed."[6] Pleasonton responding by calling the report calumny and alleging gross misrepresentations of the facts. Again, Congress refused to act.

In 1845, a new secretary of the Treasury sent two young naval officers to make a survey of the lighthouse systems of Europe. Their report, made a year later, again recommended reorganization of the Lighthouse Establishment, that it be managed by professionals with expertise in the fields of navigation and optics, that regular examinations and standardization become the norm, and that the system take advantage of the latest technological advances.

With the drumbeat call for systemic change in the management of the American lighthouse system, Congress finally acted in March 1851. As part of an appropriations bill, it was decreed that all American lighthouses be fitted with "the Fresnel system" and that the secretary of the Treasury convene a board of highly qualified military and civilian officers and engineers to "make a general detailed report and programme [*sic*] to guide legislation in extending and improving our present system of construction, illumination, inspection and superintendence" of lighthouses. The president was also required to provide officers from the engineer corps of the military to "superintend the construction and renovat[ion of] lighthouses."[7]

The board was duly appointed by the secretary of the Treasury and quickly completed its report, again reaching the same basic conclusions as previous inquiries: the system as it existed was broken, to the great detriment of the American mariners and seaborne navigation in general. Within a year a new professional management team was in place, now named the Lighthouse Board. It would oversee American lighthouses and navigational aids for the next fifty-eight years. Among the notables who served on the original Lighthouse Board and/or worked for it during this era were P. G. T. Beauregard, a West Point graduate and civil engineer who would later become a prominent Confederate general; Raphael Semmes, a naval officer who would later command the *CSS Alabama*, reputedly the most successful commerce raider in maritime history; and George Meade, also a West Point graduate and civil engineer who would, in 1863, command Union troops at the Battle of Gettysburg.

THE LIGHTHOUSE PROFESSIONALS

With the Lighthouse Board now in control, for the first time the American navigational system set out on its journey to catch up with the rest of the Western world. By the late 1850s, essentially all of the nation's towers had been retrofitted with Fresnel lenses. Lighthouses and other navigational aids were now divided into regional districts, each with a standardized hierarchy of maintenance and support. But events would soon intervene. The start of the Civil War in 1861 refocused America's now-divided loyalties. The towers that had only a few years earlier been viewed as welcoming beacons and guiding lights now became strategic assets of potential value or detriment to friend or foe depending on one's side in the conflict. For this reason, during the four years of war more than 150 lighthouses were damaged or destroyed. This unique episode in American history will be addressed in the next chapter.

Although the new Lighthouse Board made many positive changes in the entire marine navigation system, one area that had always been an area of weakness was the patronage system that played a prominent role in the placement of lighthouse keepers. American writer Ambrose Bierce, in his satirical *Devil's Dictionary*, defined a lighthouse as "a tall building on the seashore in which the government maintains a lamp and the friend of a politician."[8] As Collectors of Customs served as local superintendents of lights, "the appointment or dismissal of keepers on the grounds of political faith or heresy" was a common practice, according to an 1843 report. Thirty years later another report noted "efficient and faithful light-keepers have in many cases been changed by Collectors of Customs for no other reason than to give place to some political favorite." George Putnam, who would be appointed commissioner of lighthouses under a 1910 reform, recalled in his autobiography, "In old papers at the office I ran across books of blank forms, three to the page, check book style, with blanks for the name of the keeper addressed and the date, and informing him to the effect: 'You are superseded as keeper of _____ light station on _____ 18__, by_____.' Just this and nothing more, except for the signature of the Superintendent of Lighthouses."[9] Serious efforts were made to curb political patronage in government employment through the Civil Service acts of 1871 and 1883, but lighthouse keepers were not initially included in the system. This omission was rectified in May 1896 by a presidential proclamation issued by Grover Cleveland instructing that keepers' positions be listed with local Civil Service boards. In 1884, the Lighthouse Board sought to instill a sense of professionalism by introducing uniforms, initially for male lightkeepers as well as the crews of lightships. There were more formal versions with coat, vest, trousers, and cap, and less formal types for work. The keeper's badge was displayed on the cap.

While most lightkeepers were dedicated to their jobs irrespective of their method of appointment, the discouragement of political patronage and the efforts of the Lighthouse Board to frame these positions as careers gradually transformed the lighthouse service to one with a higher degree of long-term experience and dedication.

A NEW ERA FOR A NEW CENTURY

In 1903, the Lighthouse Board was transferred from the Treasury Department to the newly created Department of Commerce and Labor.[VII] The year 1910 brought a reorganization of how American lighthouses and related navigational aides were supervised and managed. The Lighthouse Board was abolished, and a new Bureau of Lighthouses was established. The operating name of United States Lighthouse Establishment (USLHE), which had been in use for more than a century, was changed to United States Lighthouse Service (USLHS). The Bureau and USLHS were to operate under a new commissioner of lighthouses, George Putnam.

George Rockwell Putnam was born in 1865 in Davenport, Iowa, then a relatively young and still-small city on the upper Mississippi River. Curious about the world around him, he spent much of his youth in outdoor pursuits, eventually settling on a career with the United States Coast and Geodetic Survey. His assignments during his twenty-year tenure included mapping the rugged southeastern Alaska-Canada border and traveling to Greenland and the Arctic with the famed American explorer Robert E. Peary. Following the Spanish-American War, Putnam served for several years in an administrative position in the Philippine Islands.

VII The Department of Labor would split off as a separate entity from the Department of Commerce in 1913. Lighthouse administration would remain under Commerce.

Fig. 5B George R. Putnam, U.S. Commissioner of Lighthouses, 1910–1935.

In 1910, he was asked to assume the role of commissioner for the newly created Bureau of Lighthouses. He would serve there for twenty-five years, retiring in 1935.

At the time Putnam took control of the newly named Bureau of Lighthouses, the service had expanded to a massive scale, in charge of lighthouses and navigational aids not only on America's seacoasts, but also on its navigable rivers and lakes. In the 120-odd years since the nation's founding, the number of lighthouses alone had increased more than a hundredfold. Other navigational aids that fell under the bureau's control numbered more than ten thousand. (See chart.) Under Putnam's administration, these numbers would continue to increase. By the time of his retirement in 1935, total navigational aids numbered about twenty-four thousand. However, due to Putnam's skillful stewardship and organizational efficiency, the number of bureau employees decreased between 1910 and 1935 from 5,832 to 4,980, effectively increasing the number of navigational aids from 203 to 495 per 100 employees during this time.

LIGHTHOUSE AND NAVIGATIONAL AIDS, 1791–1916[10]

	1790	1820	1850	1880	1910	1916
Lighthouses	12	59	297	661	1,397	1,706
Lightships	0	1	35	31	54	53
Lighted Buoys	0	0	0	0	225	512
Fog Signals	1	3	49	194	457	532
Bell & Whistle Buoys	0	0	0	34	267	321
Other Buoys & Minor Aids	32	186	1,155	4,301	9,261	11,823
Total Aids	45	249	1,536	5,221	11,661	14,947

Perhaps more importantly, Putnam's administration saw the increasing automation of lighthouses and related structures, the beginning of a trend that would spell the end of the traditional system by the last decade of the twentieth century. Most lighthouses had been converted to electric illumination by the 1930s. Radio beacons, with their vast geographic range, at first augmented, then supplanted the role of many light towers, at the same time lessening the importance of lightships.

THE FINAL CHAPTER

What might be described as the final chapter in the history of American lighthouses began in 1939 with the transfer—once again—of the lighthouse service from the Department of Commerce to the Treasury Department. During the Roosevelt administration and the economically dark days of the Great Depression, the president made a number of structural changes in American government "designed to increase efficiency and cut expenses." Acting under the Reorganization Act of 1939, as part of his Reorganization Plan No. II,[VIII] Roosevelt transferred Bureau of Lighthouses back to the Treasury Department to be consolidated with the Coast Guard, then also under Treasury's administration. In his message to Congress, the president justified the move as "sav[ing] money on equipment and administration and [permitting] better use of personnel."[11]

Although not mentioned in Roosevelt's text, there were other reasons for merging the lighthouse service into the Coast Guard. In Europe, clouds of war were on the horizon, and it was clear to many in Washington that the United States' eventual entry into the conflict was inevitable. As experience with the German U-boat fleet in World War I had shown, the oceans on either side of America were no longer a major barrier to foreign enemies. Less than four months later, German troops swept into western Poland, beginning the greatest conflict the world had ever known.

World War II began an era of armed conflict that would last for much of the rest of the century. Concomitantly, this was an era of technological advances across a broad range of fields including navigation. Many of these were spurred by military concerns. As noted in a previous chapter, radio-based navigation came into common use, as did radar. Early German development of ballistic missile technology would lead to the so-called "space race" of the 1950s and 1960s, all heavily dependent on the ability of an autonomous craft to determine with utmost precision its position in space and time.

VIII The name "Reorganization Plan No. II" is taken directly from Roosevelt's message to Congress. It has been copied and recopied by various authors as "Plan #11."

Military satellite-based navigation became a practical reality with the Doppler effect-based Transit system (also known as NAVSAT for Navigation Satellite System) in the mid-1960s. By the late 1970s, location was being determined by observed differences in the timing of radio signals received from multiple satellites, termed the global positioning system (GPS). By the 1990s, GPS was commonly in use in both military and civilian applications. As the system was owned and operated by the United States military, during the initial years of the public's use of GPS, the signal was deliberately degraded through distortion of the time signals transmitted from the system's fleet of satellites, a policy known as Selective Availability (SA). The stated reasoning for this was based on the fact that a highly precise GPS could be used to guide ballistic missiles. As such, during these last days of the Cold War, it could represent a potential military threat to the nation. Even with SA, however, location could be determined with an accuracy of one hundred meters of a receiver's location.[IX]

The use of Selective Availability came to an end on May 1, 2000, when President Clinton issued an order to deactivate this feature, allowing GPS users to determine their location with an accuracy of approximately nine meters. This change opened an entire new world of possibilities for inexpensive, highly precise location applications, ranging from smartphones to automobiles to agriculture. In the two decades since SA was discontinued, the reliability and accuracy of the GPS system has increased dramatically. As of 2018, mean position error determinations were less than two meters.[12] By locking on to multiple satellites and taking advantage of earth-based augmentations services, location can now be determined with an accuracy falling within a few centimeters of a given point. An interesting, if not ironic, observation is that the same basic principle that allowed Thomas Harrison's chronometers to estimate location within tens of miles in the 1730s is now being applied to determine location within tens of centimeters.

[IX] In other words, a person with a good GPS receiver in the 1990s would be able to say with accuracy that he or she was located somewhere within a circle roughly 660 feet in diameter.

With the exception of the historic Boston light, by the end of the twentieth century all of America's lighthouses had been decommissioned or automated. Many were offered to conservation or historic preservation groups to be maintained as relics of the past. The last civilian lighthouse keeper, Frank Schubert, died at age eighty-eight on December 11, 2003. Even though his charge, the Coney Island lighthouse in Brooklyn, New York, had been automated in the early 1990s, he was allowed to live out his final days in the seven-room keeper's cottage beneath the 1890 iron and steel tower.[13]

CHAPTER 6

LIGHTHOUSES DURING THE AMERICAN CIVIL WAR

The Civil War of 1861 to 1865 was a horrific conflict, tearing asunder the fragile fabric of the American republic while leaving lasting scars and regional animosity that persist more than a century and a half later. One of the root causes of the war was the institution of slavery and, implicitly, the Southern way of life built upon this foundation. The economics of slavery, more so than a debate on its morality, were central to the conflicts that led to secession and the struggle that followed. The labor of slaves had become an integral part of the production of cotton, the single crop that defined the Southern economy and whose export to foreign markets provided a positive balance of trade for the relatively young nation. In the mid-nineteenth century, the revenue of the federal government was highly dependent on import and export duties. In the brisk transatlantic trade that developed after Eli Whitney's cotton gin made mass production of cotton possible, cotton exports to England and France and the returning harvest of European goods to be sold on the American market yielded major streams of revenue for the support of essential central governmental functions. One of the reasons the states of the South felt reasonably secure in their decision to secede from the Union was the relative economic strength of the region. As one contemporary author noted, during the prewar period, "four states alone, Virginia, the two Carolinas and Georgia, defrayed three-fourths of the expenses of the General

Government."[1,1] In 1860, Georgia was reputed to be the wealthiest state in the nation but with the caveat that half of this wealth was based on the value of slaves.

As a significant portion of the income of the United States government was based on import and export duties, the Commissioners of Customs occupied an important role in funding governmental operations. Prior to the creation of the Lighthouse Board, these officials were also in charge of lighthouses, indicating the important role these structures played in the intertwined fields of navigation and commerce. When the seceding states met in Montgomery, Alabama, in the winter of 1861 to craft a constitution and the laws of the Confederacy, this role was not forgotten. Many of the founding documents of the Confederate States of America closely resembled those of the United States but with changes designed to reflect the ethos and politics of the breakaway nation. Continuation of the institution of slavery was affirmed, as was the relative independent role of the individual states. The president and vice president were limited to a single six-year term. Monetary appropriations required a two-thirds approval of both houses of the new Congress. On a practical basis, most existing governmental institutions were kept in place; simply the name was changed. It was assumed, for example, that the United States Postal Service would become the Confederate States Postal Service.

The eleven states that would make up the Confederate States of America did not secede in unison. It was a painful choice for many, and the conventions that would make this monumental decision were venues of lively debate. South Carolina was first, seceding on December 20, 1860, and North Carolina and Tennessee were last, seceding in May and June 1861, respectively. As their states left the Union, many congressmen, senators, military officers, and others holding positions of power, trust, and skill within the federal government did so as well, casting their lots in with those

---

[1] The proportional contribution of the Southern cotton-producing states to the federal government's revenue in this era has been variously described as between two-thirds and three-quarters of the total. While there is some variability, the basic fact remains that the South in those days was the United States' economic powerhouse.

of their states, and, in turn, with the Confederacy. One such man was Raphael Semmes.

Semmes was born in Maryland and spent most of his life in naval service. He enlisted as a midshipman in the United States Navy at age seventeen in 1826 and rose steadily through the ranks. In 1841, Semmes and his family moved to Alabama, which he considered thereafter his adopted home state. Semmes served with distinction in the Mexican-American War of the late 1840s, and by 1855 had been elevated to the rank of commander. During the latter part of that decade he was assigned to the newly created Lighthouse Board, later assuming the office of board secretary.

Fig. 6A Raphael Semmes

By 1860, it was clear to many that several Southern states were actively planning secession. Confiding in members of the Alabama congressional delegation in Washington, Semmes stated that if Alabama seceded from the Union, he was willing to resign his naval commission and pledge his allegiance to the cause of the South. In February 1861, a month after Alabama's formal decision to secede, Semmes received a summons via telegram from the chairman of the Committee on Naval Affairs of the Provisional Confederate Congress. He resigned his commission the following morning and left for Montgomery, where Congress was in session.

Over the following days Semmes met with the provisional president, Jefferson Davis, and with congressional committees on military and naval affairs. Given his experience in both the management of lighthouses and in naval warfare, he advised members of Congress on the establishment of a lighthouse board similar to that of the United States government, and on which type of navy the Confederacy—with its limited resources—would find most effective. Regarding the latter, Semmes suggested "an irregular naval force," specifically "a well-organized system of private armed ships, called privateers." In case of conflict, the damage that the South could inflict on the North from this type of irregular warfare would be economic. "If you are warred upon at all," Semmes wrote, "it will be by a commercial people whose ability to do you harm will consist chiefly in ships and shipping. It is at ships and shipping therefore you must strike, and the most effectual way to do this is by means of the irregular force of which I speak."[2] As war clouds gathered, Semmes was sent north to acquire armament and related military equipment for the Confederacy.

On March 6, 1861, the Provisional Congress in Montgomery formally passed "An Act to Establish a Bureau in Connection with The Department of the Treasury, to be Known as The Light House Bureau." It was closely modeled on the equivalent organization of the United States government and was given control of "all light houses, light vessels, buoys, and other aids to navigation." As events would play out, however, the concept of a fully functioning Lighthouse Bureau would never become a reality.

One of the major issues accompanying secession was the fate of assets in the seceding states that had formerly belonged to the federal government, including military installations and lighthouses. The flash point came in South Carolina, where state authorities demanded that the US Army abandon its bases on the approach to Charleston Harbor, specifically Fort Moultrie and Fort Sumter. Seeking a defensible position in case of attack, the local commander, Major Robert Anderson, consolidated his forces in Fort Sumter, a more substantial and well-armed redoubt. President James Buchanan, serving out the last days of his term, attempted to reinforce and resupply the garrison in early January 1861, but the supply ship was driven away by cannon fire from shore batteries manned by the South Carolina militia. After several failed attempts at compromise over the next three months, the shore batteries began a sustained artillery bombardment of Fort Sumter on April 12th. The fort formally surrendered on April 14th, marking the beginning of hostilities in a war that would claim more than six hundred thousand American lives.

In response, on April 16th newly inaugurated President Abraham Lincoln issued a proclamation declaring the inhabitants of the states of Georgia, South Carolina, Virginia, North Carolina,

BY THE PRESIDENT OF THE UNITED STATES OF AMERICA.
A PROCLAMATION.

Whereas, for the reasons assigned in my Proclamation of the 19th instant, a blockade of the ports of the States of South Carolina, Georgia, Florida, Alabama, Louisiana, Mississippi, and Texas was ordered to be established;

And whereas, since that date, public property of the United States has been seized, the collection of the revenue obstructed, and duly commissioned officers of the United States, while engaged in executing the orders of their superiors, have been arrested and held in custody as prisoners, or have been impeded in the discharge of their official duties without due legal process, by persons claiming to act under authorities of the States of Virginia and North Carolina :

An efficient blockade of the ports of those States will also be established.

In witness whereof, I have hereunto set my hand, and caused the seal of the United States to be affixed.

Done at the City of Washington, this twenty-seventh day of April, in the year of our Lord [L.S.] one thousand eight hundred and sixty-one, and of the Independence of the United States the eighty-fifth.

ABRAHAM LINCOLN.

By the President:
  WILLIAM H. SEWARD,
    Secretary of State.

Fig. 6B Lincoln's proclamation of the Southern blockade.

Tennessee, Alabama, Louisiana, Texas, Arkansas, Mississippi, and Florida to be "in a state of insurrection against the United States, and that all commercial intercourse between the same and the inhabitants thereof...and the citizens of other states and other parts of the United States is unlawful...." Three days later, the president issued a second proclamation, announcing a blockade of the ports of South Carolina, Georgia, Florida, Alabama, Mississippi, Louisiana, and Texas, which he extended to Virginia and North Carolina by a third proclamation on April 27th. The purpose of the blockade was to prevent the export of the cotton, the South's main source of revenue, and to block the import of arms and other war matériel into the states of the Confederacy.

The blockade was part of a greater plan formulated by Winfield Scott, commanding general of the United States Army. Scott, then seventy-four years of age, had served as head of the

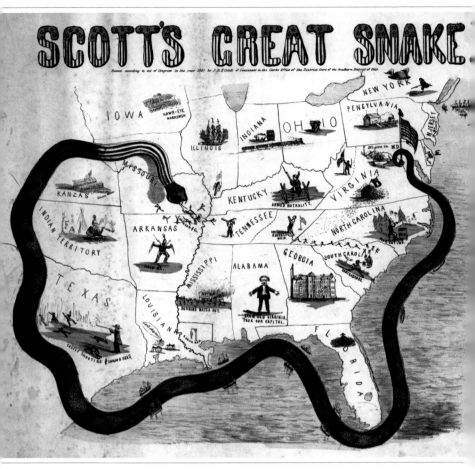

Fig. 6C General Winfield Scott's "Anaconda Plan."

nation's land forces since 1841. In order to defeat the breakaway Southern states, he recommended a two-pronged approach, one primarily economic and the other military. First, a naval blockade of the coast from Virginia to Texas would restrict trade and cut off the Confederacy's access to money and arms. Second, military intervention would proceed via push down the Mississippi River, cutting the Confederacy in half and capturing the major ports of New Orleans and Mobile. Afterwards, conventional thrusts by troop columns would complete the military takeover. Although never fully implemented as initially envisioned, the scheme caught the imagination of the public and was termed the "Anaconda Plan," as a visual depiction of the blockade was said to resemble an anaconda strangling its prey.

At the outbreak of hostilities, the American navigational system was at its historical peak. By 1859, the new Fresnel lenses had been installed in every major lighthouse. The treacherous waters along the coasts of the Florida peninsula were well marked by powerful beacons to help guide ships safely to and from the major Gulf ports of Mobile and New Orleans. Proper maintenance and a ready supply of oil for illumination enabled busy lanes of seaborne commerce. Now, the implementation of the blockade of Southern ports radically changed the role of lighthouses along the coasts of the Confederacy. The ships enforcing the blockade would be in unfamiliar waters, and, accordingly, see lighthouses as navigational assets. On the other side, the Confederates, relying on blockade runners and privateers in ships manned by sailors familiar with local waters, would prefer not to offer the enemy any advantage that might assist in their detection and capture as they attempted to stealthily enter and exit coastal waters.

The Confederate States Lighthouse Bureau had been created with the assumption that the basic system as it existed under federal administration would continue after secession. Raphael Semmes, who might have led the bureau had the transition been peaceful, was soon himself acting as a privateer in command of the commerce raider *CSS Sumter*. Commander Ebenezer Farrand, a former US naval officer, was appointed bureau chief on May 1st, moving with the rest of the government to the Confederacy's new capital of Richmond, Virginia, in the months that followed. Initially, little seemed to change. A good example is the salary payment receipt for George Davis, an assistant keeper of the Tybee light near Savannah. The form on which Davis made his mark—it is assumed that he was illiterate—had the printed heading of "U.S. Light-House Establishment." On the form, dated March 31, 1861, the "U.S." on the heading had been scratched through, and written over are the words "Confederate States." (See Figure 6D.)

The United States Lighthouse Board was charged with maintaining the navigational system of the nation's waterways. In sharp contrast, the Confederate States Lighthouse Bureau presided over the deliberate extinguishing and oftentimes destruction of the

South's coastal lighthouses.ⁱⁱ Shortly after the blockade began, local activists—sometimes Collectors of the Customs, lighthouse keepers, or simply groups of individuals who felt the lights provided more aid to the blockaders than to the Confederate cause—began extinguishing lighthouses. The Fresnel lenses were expensive and of European manufacture. In many cases, they were dismantled and hidden, hopefully to be recovered and reinstalled at the end of hostilities. By September of 1861, essentially all of the coastal lighthouses of the Confederate States had been extinguished, with the exception of seven relatively inaccessible lights in the Florida

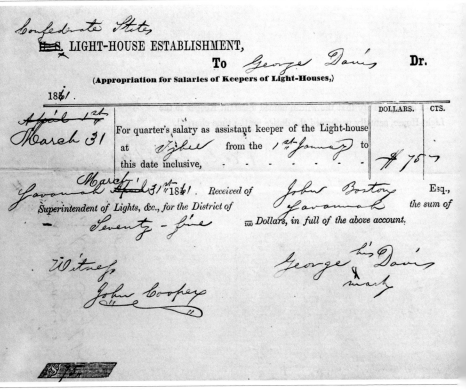

Fig. 6D A receipt from the newly inaugurated "Confederate States Light-House Establishment," dated March 31, 1861. Note that the assistant keeper's salary for three months was seventy-five dollars.

II For a more comprehensive discussion of the Confederate States Lighthouse Bureau, see the article written by David Cipra for the United States Lighthouse Society: https://uslhs.org/confederate-states-lighthouse-bureau-david-cipra.

Keys. While the Confederate Lighthouse Bureau continued to exist throughout the conflict, its purpose and ability to make a meaningful difference in the fate of its charges was secondary to the exigencies of armed conflict. Most of the naval personnel initially appointed to managerial positions in the bureau would soon leave to undertake other duties for the Confederate Navy.

A brief sampling of some of the news reports from the fall of 1861 lends insight into the fate of various Southern lighthouses, as well as the attitudes of the day toward the opposing side. *The Augusta* (GA) *Chronicle* of September 12, quoting the *Petersburg* (VA) *Express*, reported under the headline "Gallant Exploit of Confederate Troops":

A lighthouse, located at an important point on the Back River in Elizabeth City County [Virginia], about six miles from the Old Point and known as Back River Light, having furnished considerable aid to the Federal pirates who have been navigating in that section for some time past, it was determined by Confederate authorities at Yorktown to extinguish the prominent illuminator. In furtherance of this object, Lieut. John A. Dickson, of the Burke County (N.C.) Rifles, accompanied by twenty men of his company and twenty others of the Buncombe (N.C.) Rifles, left Yorktown last Tuesday afternoon at 4 o'clock in three boats, and arrived at their point of destination about 11 o'clock at night. They disembarked some three miles from the lighthouse, and having secured their boats and procured the aid of some good and loyal militia men were safely guided to the spot. The house was surrounded, a keeper, a man named Hawkins, secured, and then the lamps and building totally demolished. The keeper's wife and children were treated with the utmost kindness and consideration, but Hawkins, the Southern traitor and Lincoln office holder, was taken to Yorktown.... The destruction of the light, we are informed, seriously interferes with the navigation of the Chesapeake Bay, and is hoped may contribute to the loss of several of Lincoln's piratical craft.[3]

The next day's issue of the *Chronicle* quoted the *Tampa Peninsula* under news of "The Florida Blockade":

Mr. Rickards, the light-keeper, arrived at this place on Saturday last bringing the intelligence that the blockading steamer took her departure from our waters on Thursday of last week. The Lincolnites, before they left, were driven to the necessity of burning the cutter Appleton, tender to the steamer, she having been blown high and dry on Egmont Key by the late blow. It is more than probable that the absence of the steamer is only temporary—her object, doubtless, being to procure another tender. Apropos—on receipt of the above intelligence, detachments of the Sunny South Guard, Coast Guards and a number of our citizens, proceeded to Egmont Key and removed the lamps, oil, etc. from the lighthouse and brought them to Tampa. These lamps will never again give light to the benighted followers of King Abe.[4]

An item published during the fall of 1861 in New York's *Frank Leslie's Illustrated Newspaper* reflects the Northern view of the Confederates' destruction of lighthouses: "Soon after the bombardment of Fort Sumter, the Confederate Government, with that murderous indifference to human life which has distinguished them from the first, extinguished all the lights they could reach, and among others the lighthouse erected at Cape Hatteras."[5]

Although military units were often involved in lighthouse raids, on several occasions, ardent civilian supporters of the Confederacy took it upon themselves to extinguish these navigational lights. The fate of two south Florida lighthouses provides a good example. In 1861, Florida was a sparsely populated backwater, having been acquired from Spain in 1821 and admitted to statehood only in 1845. Most of the population lived in coastal settlements, and most of these were on the northern Atlantic and Gulf coasts. According to a contemporary report, the southern Atlantic coastal region "can hardly be said to be inhabited, and is of no great consequence except as a convenient place of resort for pirates." Fewer than three hundred people lived on the Atlantic coast south of Cape Canaveral, mostly in isolated settlements on a narrow strip of land between the sea to the east and swampland to the west.[6]

While it may not have been a center of population and commerce, Florida's south Atlantic coast occupied a vital position in relation to sea routes. The deep waters of the Florida Straits and Gulf Stream just offshore lay perilously near to submerged coral reefs in a region prone to sudden violent storms. In response, the US government had erected a lighthouse in 1825 at Cape Florida on the southern entrance to Biscayne Bay, and another completed in 1860 at Jupiter Inlet, just north of modern Palm Beach.[III] By midsummer of 1861, some four months into the Federal blockade, both lighthouses remained intact and functional.

At the Jupiter Inlet light, assistant keeper August Lang, a German immigrant, encouraged head keeper Joseph Papy to follow other keepers who had extinguished their lights in the presence of the blockade. Papy, although expressing loyalty to the Confederacy, refused, citing an absence of formal orders to do so. Lang resigned and sought out James Paine, "a settler of strong Confederate sympathies," who lived some forty miles to the north.[7] A few days later, Lang and Paine returned to the lighthouse with the purpose of rendering it inoperable. When Papy demanded to know under what authority they were acting, Paine replied, "We came as citizens of the Confederate States, to discharge a duty to our country."[8] The two then commenced to dismantle the lenses, prisms, and rotating mechanism of the light.

Emboldened by their success in disabling the Jupiter Inlet light, the two men considered the possibility of extinguishing the Cape Florida light, the sole remaining beacon along the south coast. While Paine stayed behind, Lang and three other men set out on the arduous ninety-mile journey south. Arriving approximately a week later, they were informed that the keepers were armed and ready to use force to repel anyone seeking to damage the lighthouse. When they approached the lighthouse under cover of darkness to assess the situation, they found the keepers had locked themselves inside the near-impregnable tower behind an iron door. As a ruse, they convinced the keepers that were delivering supplies from Key West. When the door was opened, the raiders took control of the lighthouse, smashing the lenses and reflectors and stealing several lamps as well as firearms belonging to the keepers.

In most cases, it was impossible to prevent many beacons from being disabled, primarily due to the number of lighthouses and the limited resources available to Federal forces. In response

---

III The Jupiter Inlet Light was designed by and built under the direction of then-Lieutenant George Mead, who would later be the commanding general of Union forces at the Battle of Gettysburg. A more detailed description of the attacks on the Jupiter Inlet and Cape Florida lighthouses can be found in Neil E. Hurley's *Florida's Lighthouses in the Civil War* (Middle River Press, 2007).

Fig. 6E The remains of the lighthouse at Fort Morgan on Mobile Bay, Alabama, after the Union bombardment in 1864.

to the extinguishing of the Jupiter Inlet and Cape Florida lights, a Union officer assigned to blockade duty complained in a letter to a friend, "Where is the Navy? And what are they doing? Nothing!"[9] A Northern newspaper commented, "The Cape Florida lighthouse has been blown up [sic] by the rebels, who have shown the spirit of Vandals in destroying these guides to commerce on the Southern coast, built not for the benefit of the North alone, but of the whole world."[10] The only lighthouses in the Confederacy that remained under Northern control (and illuminated) during the war years were those located on or near the Florida Keys, an area protected by both the sea and a strong presence of Union forces.[IV]

The exact dates of the disabling of the five lighthouses that survive today on the Georgia coast is a bit uncertain. At Tybee light, there is a June 1861 invoice in the National Archives submitted to the "Confederate States Light House Department" for removal of the tower's lens. (See Chapter 10.) At Cockspur, the light was removed in February 1862 as Federal troops occupied Tybee Island and prepared to attack Fort Pulaski. The lighting apparatus from the Sapelo lighthouse was said to be stored in Macon in December 1861, apparently having been removed earlier in the year. The third-order Fresnel lens had been removed from the St. Simons lighthouse and stored in Brunswick by the time the tower was destroyed by Confederate troops in September 1861. The Little Cumberland light was removed and stored during this same time frame, but its whereabouts and eventual fate both then and now are unknown. Although there were searches by Federal troops seeking these valuable lights during and after the conflict, none were recovered.

As the North took control of disabled lighthouses, efforts were made to reestablish function for the ones most important to navigation. A news item from early January 1863 reported, "The schooner *Pharos*, which sailed from New Bedford [Massachusetts]... on her way South carries out material for the reconstruction of lighthouses destroyed by the Confederate vandals,"[11] but even then Confederate forces were often able to do damage through late in the conflict. In April 1864, for example, "forty rebels landed on Cape Lookout [on North Carolina's Outer Banks], secured the lighthouse keepers, and exploded a keg of powder in the lighthouse, injuring it greatly."[12] In early November of the same year, it was reported that "the ram *Albemarle* ran out the other night in the sound and on the 4th instant reached the Croatan lighthouse, the keeper of which was captured, the lighthouse blown up and its contents destroyed."[13] The Croatan light was a then-new offshore screwpile light that had replaced a lightship just prior to the outbreak of the war.

[IV] These included two tower lights on the Dry Tortugas and one at Key West, plus four screwpile wave-swept lights at Carysfort Reef, Sombrero Key, Sand Key, and the Northwest Passage.

As the nation began the long process of rebuilding after the cessation of hostilities, the *New York Herald* noted,

> The Lighthouse Board is constantly directing the reestablishment along the Southern coast of lighthouses destroyed by the rebels during the late war. Large appropriations will be required to put the coast in the condition it enjoyed prior to 1861. Nearly every lighthouse from Cape Henry [Virginia] down the Atlantic and Gulf coasts with the exception of a few in Florida, were destroyed by the enemy, the structures being torn down and the lenses broken or carried away.[14]

A few lighthouses were damaged during artillery bombardments. The Cockspur light on the Savannah River, for example, received only minor damage while the lighthouse at Fort Morgan on Mobile Bay was nearly completely destroyed. Overall, the degree of destruction had indeed been massive, involving more than 150 lighthouses, as well as buoys and other navigational aids. The massive task of returning these lighthouses and aids to full operational status was not completed until 1875, a full decade after the end of the war.

CHAPTER 7

KEEPERS OF THE LIGHT

Perhaps not surprisingly, the perceptions of mystery, romance, and symbolism that lighthouses have come to represent extend to their keepers as well. Often depicted as stalwart men (and occasionally women) who chose to forsake life's mainstream for a monastic existence in remote seaside towers, lighthouse keepers have been imbued with an aura of courageous self-sacrifice and dedication to the good of mankind. Words used to describe the life of a lightkeeper commonly include "monotonous" and "lonely," but for many the solitude was broken by the moods of the wind and sea and tempered by visits from friends and family. In the words of a writer describing the life of a fictional keeper in 1867,

Our duties at the lighthouse were not very onerous, but of course they had to be performed with the order and regularity of machinery, the first article in the regulations issued to lighthouse keepers running as follows: "You are to light the lamps every evening at sun-setting, and keep them constantly burning bright and clear until sun-rising." Whatever may happen, come fair or foul weather, as sure as the sunset, is the beacon-light to shine across the trackless sea, a warning to sailors of some treacherous and shifting sand bank. The monotony of our life was broken only by storms and visits from our friends on land. It was trying to a man's nerves to sit in the watch-room, immediately below the lantern, during a gale. The waves seemed to leap in anger against the light, which steadily shown to warn ships against the lurking rock; the ocean dashed against the shore with reverberating thunder, and our stout wooden beams and rafters, dovetailed and clamped together by massive iron bands, rattled and shook as if the next moment would see the whole fabric whirling in the angry sea. A strong contrast to such nights were the calm summer evenings, when the ocean stretched round us for miles, dotted here and there with white sails, and troubled only by a passing shoal of mackerel or a puff of the summer breeze.[1]

Reality, however, was often far more prosaic. One historian noted,

Although keepers are pictured as belonging to an almost heroic breed, they were not always so highly thought of in the United States, particularly in the early days. Too many of the light keepers looked upon their jobs as sinecures on which not a great deal of energy need be spent. This attitude was engendered by the fact that the position of light keeper was often subject to political appointment.[2]

While it is true that in America's early days many, perhaps most, lighthouses were located in remote and sparsely populated areas, over the decades the relentless march of urbanization overtook them, turning many of the once-isolated towers into readily accessible tourist attractions.[1]

AMERICAN LIGHTHOUSE ERAS

Broadly speaking, the quality, experience, and professionalism of lighthouse keepers in the United States can be divided into three eras. The first encompasses the years from 1789 and the federal assumption of ownership and administration of lighthouses to 1852 and the formation of the Lighthouse Board. In his preface to *America's Lighthouses*, Holland notes,

If one had to indicate the single most important year in the history of this nation's aids to navigation, it would unquestionably have to be 1852, the year the Lighthouse Board came into existence. At that time the United States' lighthouses and other aids to navigation taken as a whole gave this country a third-rate lighthouse service. The lighting system that illu-

---

[1] In Georgia, for example, the lights on Tybee, Cockspur, and St. Simons Islands, once far from centers of population and commerce, are now well within the metropolitan areas of nearby cities. Lighthouses on Sapelo and Little Cumberland Islands, in contrast, remain isolated, attracting relatively few visitors.

minated the lighthouse towers was not only out of date, but in its best days did not come up to the quality of that of other nations with similar systems. Many towers had been poorly constructed and many aids to navigation had not been placed so as to be of maximum advantage to the mariner. The nation's coasts were not adequately marked, lights were poorly maintained by poorly trained keepers, many of whom were political hacks. Moreover, the administration of aids to navigation could be termed, at best, inept.[3]

The second era includes the years from 1852 to 1939. During this time the American navigational system evolved and expanded with a vast increase in the number of lighthouses and other aids to navigation, as well as the routine integration of new technologies. Concomitantly, the role of lighthouse keepers transitioned from that of simple employment to profession as the nation's system grew to be among the finest in the world. Patronage remained a problem until the inclusion of keepers in the Civil Service system in 1896. Prior to this, politicians and Collectors of Customs often viewed these jobs as rewards to be dispensed to party faithfuls and supporters, with actual knowledge and experience being secondary considerations. During Abraham Lincoln's first administration, for example, about three-quarters of keepers were replaced by political appointees.[4]

The third era began with the abolishment of the Bureau of Lighthouses in 1939 and the transfer of administration of all navigational aids to the Coast Guard. The decades that followed witnessed the practical end of the lighthouse era. By the conclusion of the twentieth century, essentially all of America's lighthouses had been automated and marginalized, their roles supplanted by the use of new alternative methods of location-finding.

DUTIES AND UNIFORMS

A lighthouse keeper had undertaken a demanding, twenty-four-hours-a-day, seven-days-a-week job requiring commitment and stamina. Performed properly, a keeper's duty began near sunset with the lighting of the lamps, continued through the night to

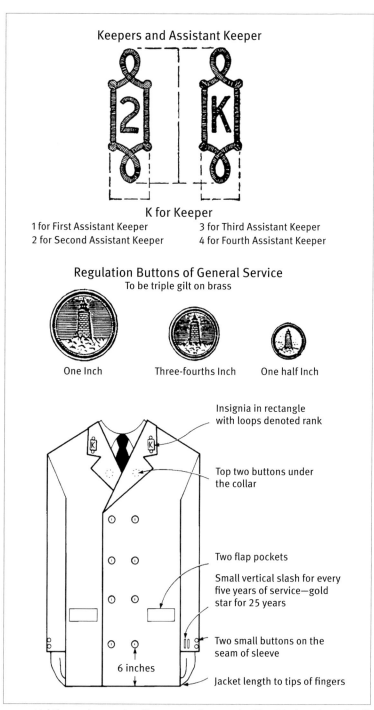

Fig. 7A Lighthouse keepers' uniforms were first required in 1884 in order to instill a sense of professionalism.

ensure proper functioning of the light, trimming of wicks, winding the mechanism that turned the rotating beam, etc., and ended in the late morning of the following day with the cleaning of the apparatus in preparation for the process to begin once more that evening. In the early days, training was on the job and empiric. Under the new administration of the Lighthouse Board, much of this changed. The board began to organize and specify the duties of keepers, with detailed written instructions as to what should be done and by whom, and how and when it needed to be done. For lighthouses with assistant keepers—a more common pattern of staffing with the passing years—each keeper was assigned a set of daily tasks. These included not only maintenance of the light and tower itself, but also associated navigational aids such as foghorns, range lights, and nearby buoys. In addition to these instructions, the board required a greater level of documentation by keepers, including daily event logs and maintenance records. Even while the patronage system of choosing keepers remained in place, the board instituted a uniform system of monitoring, discipline, and punishment (or termination) to assure the consistent functioning of light stations and related navigational aids.

In 1883, the Lighthouse Board mandated the wearing of uniforms for lightkeepers and other personnel, stating, "It is believed that uniforming the personnel of the service, some 1,600 in number, will aid in maintaining its discipline, increase its efficiency, raise its tone and add to the esprit de corps."[5] Effective May 1, 1884, male keepers, assistants, masters and crews of lightships, and other support staff were required to wear "at all times on duty" a double-breasted navy-blue coat, pants, and brimmed cap adorned with brass buttons and various insignia indicating rank and tenure. Less formal garb was specified for work assignments. Female keepers and assistants—relatively few in number—were not required to wear uniforms. The Civil Service paid for the initial outfitting; thereafter, keepers were required to buy their own. Uniforms, with various changes over the years, remained a requirement through the end of the twentieth century.

The salary of a lighthouse keeper was modest, even by current standards.[6] In 1840, an average keeper might be paid four hundred dollars per year. In 1867, Congress increased keepers' pay to six hundred dollars per year, a figure that stayed the same until 1918 when it was increased again to an average of $840 per year. By the 1930s, keepers were earning a yearly average between one thousand and two thousand dollars. While appearing to be scant pay for demanding work, there were other benefits. Housing was generally provided for the keeper and his family. Extra pay was frequently awarded for remote or otherwise less desirable locations. The yearly renumeration might also include fuel for heating, or other forms of support or sustenance. Some lighthouses had adjacent or nearby land available where a keeper and his family might raise a garden or even keep a few chickens, goats, cattle, or other animals. In some situations, the keeper in his spare time might supplement his income with fishing or other limited work if available. With the transfer of lighthouse administration to the Coast Guard in 1939, keepers were given the choice of remaining as civilian employees or transferring to military service with no loss of pay.

THE LIFE OF A LIGHTKEEPER

Most keepers were men, and many had wives and children. The Lighthouse Service preferred to hire married men, believing they would be more content with their otherwise lonely jobs. In some situations, the keeper's family might live ashore, a short commute by boat. For many others, though, the lighthouse was part of a complex that included the light tower and quarters for the keeper, assistant keeper(s), and their families. Lightkeepers and those who married them were most often self-sufficient folk, aware of the rigors and demands of the situations they chose. The keeper's wife and older children frequently became de facto assistant keepers, sometimes formally hired in that position, but more commonly as informal assistants to their husbands and fathers. Historical lighthouse accounts abound with stories of the commitment and perseverance of keepers' wives and children, often taking over management of the light station in times of his illness or absence. With children, there were concerns about schooling and education, with each set of circumstances being different. In some situations, it was

Fig. 7B The "The Lighthouse Keeper's Daughter," by Norman Rockwell, 1923.

possible for the children to commute to a nearby school, and in others they would stay on the mainland during the week.

As a further effort to maintain the morale of isolated keepers and their families, in the mid-1870s the Lighthouse Service began providing rotating collections of books to some light stations. These small libraries consisted of forty or fifty books contained in a sturdy wooden case designed to be easily transported from station to station. "It is intended that each library remain about six months at a place, when it will be exchanged for another. By this means the keepers will be constantly supplied with fresh and interesting reading matter and be made more contented with the lonely life and routine duties of these distant and often inaccessible stations," the Lighthouse Board's annual report for 1876 stated. The libraries

Fig. 7C A "Keepers' Library" on display at Tybee Light Station. These contained an assortment of books designed to help relieve keepers' monotony.

contained an assortment of books ranging from religious topics to novels and works of nonfiction. Each would stay at a station for a few months before being exchanged for another library with a different assortment of books. By the 1880s, there were more than four hundred mini-libraries in circulation. The introduction of radio and other forms of wireless communication in the 1920s lessened the appeal of these libraries, but some remained in circulation well into the mid-twentieth century.[7]

LIGHTHOUSE HEROES

In the minds of many, the archetypal lighthouse keeper is pictured as an older white male with greying hair and beard, peering intently out to sea while dressed in his brass-buttoned blue uniform. While correct in some cases, keepers were a more diverse lot. While black and other minority keepers were the exception, careful searches of existing records have revealed a significant number in the coastal states of the former Confederacy.⁸ Female keepers and assistants were the exception as well yet played a prominent role in the popular history of American lighthouses, as will be noted below. Most were the wives or daughters of male keepers, reared in the profession and taking over their husband's or father's position upon his death or disability. Describing them as "modern heroines of real life romance," a Boston newspaper in 1897 noted that there were "no less than thirty women lighthouse keepers in the employment of the United States today."⁹ Based on records from the National Archives and published in *Women Who Kept the Lights*, Mary L. Clifford and J. Candace Clifford were able to document at least 142 women who served as principal keepers for at least a year between 1776 and 1939, when lighthouse administration was transferred to the Coast Guard.¹⁰

Although keepers' assigned duties were nominally those of maintaining their light stations and related navigational aids, they were also expected to render assistance to those who had fallen afoul of the sea. Lighthouse histories are replete with accounts of keepers who risked their own lives to save those of others. A sampling of newspaper reports from the latter third of the nineteenth century is typical: in 1869, Edward Pope, keeper of the Ellis Bay light in the Gulf of St. Lawrence, and his family were stricken with typhoid just as the revolving apparatus of the light broke. With no way to communicate with the shore, he rotated the light by hand every minute and a half from 7:00 P.M. till 6:00 A.M. nightly for months while attending to his family's illness during the day. It was speculated that in doing so he "may have saved a thousand lives."¹¹ In April 1886, the crews of two steamers caught in "a terrible gale" were rescued with no loss of life "through the heroic efforts of the light keeper" of the Pelee Island Light on Lake Erie.¹² In March 1890, near Norfolk, Virginia, five men of an oyster boat were lost despite the efforts of assistant lighthouse keeper James Hurst, who himself was saved from the sea by a passing boat.¹³ The following year, the keeper of the Lime Point lighthouse on San Francisco Bay rescued two men as they were drifting out to sea on their capsized boat.¹⁴, ᴵᴵ In November 1900, Ole Anderson, the keeper of the Plum Island light on the entry to New York's Long Island Sound, was

---

ᴵᴵ Lime Point is today the northern terminus of the Golden Gate Bridge.

abandoned by his assistant keeper (who was later assumed to have drowned on his way to shore). Maintaining the light was a full-time job for two men. Running short of provisions, hungry, and having to work both day and night, Anderson kept the light burning for ten days before his plight was discovered and relief arrived.[15] While these tales of perseverance and personal heroism by male lightkeepers are common, it is a bit ironic that—given their relatively small numbers—many of the best-remembered lighthouse heroes are female.

LIGHTHOUSE HEROINES

The first woman to achieve international fame due to her association with a lighthouse was Grace Darling.[16] Born in County Northumberland in northeast England in 1815, she was the daughter, granddaughter, and sister of lightkeepers employed by Trinity House, the English lighthouse establishment. When Darling was only a few months old she moved with her family to the Farne Islands, a series of treacherous rocky islets a few miles off the Northumberland coast. She would live there for the next two decades as her father kept the lighthouse on Brownsman Island. Described as a "shy and thoughtful child," she learned the ways of the sea and, like many family members of lightkeepers, assisted her father in his daily routine. In 1826, Grace and her family moved to nearby Longstone Island, where her father was to become the keeper of a newly constructed light.

At 6:30 P.M. on September 5, 1838, the *SS Forfarshire*, a twin-side-paddlewheel steamship, 132 feet long, displacing 450 tons and powered by two 90-horsepower steam engines, left Hull in East Yorkshire bound for the Scottish port of Dundee on its regular weekly run carrying passengers and freight. The ship was only four years old, and in these early days of steam-powered propulsion, had two sailing masts with full rigging in case of failure of the engines. While docked in Hull, repairs had been made on one of the boilers, but all was well on her departure north. During the night, more problems developed with the boilers. There were concerns about the safety of the ship, but Captain John Humble assured the passengers that there was no need to put in to a coastal port for repairs.

By midday the following day, the engines were unable to generate enough power to make headway, forcing the crew to raise sails. As midnight approached, the *Forfarshire* began to be battered by gale-force winds, followed shortly thereafter by the failure of both engines. Without means of propulsion and driven southward by the storm, the captain made the decision to seek shelter near the Farne Islands. Within hours, Humble spied a light that he thought was on Inner Farne Island, marking the way to a safe anchorage in which to ride out the storm. He was mistaken. It was instead the Longstone light, about two and a half miles further out to sea and surrounded by dangerous rocky islets and shoals.

At approximately 4:00 A.M., the *Forfarshire* struck and lodged upon a small rocky outcropping known as Harker's Rock, about a mile from the Longstone light. Eight members of the crew lowered a lifeboat and, accompanied by one passenger who managed to leap in, made their way to the relative safety of the open sea. Shortly thereafter a wave thrust the ship further on the rocks, causing her to break in two sections. The aft part, containing Captain Humble, his wife, and most of the passengers immediately disappeared into the stormy waters, leaving the bow section and a few survivors on the rock. Here Grace Darling enters the story.

Looking out to sea at first light shortly before 5:00 A.M. on the morning of the 7th, Grace spotted the remains of the stranded ship in the direction of Harker's Rock. She alerted her father, and together they studied the scene through a telescope. Initially, no survivors were seen, but as the light improved, they were able to make out three or four people on the wave-swept rock. As lighthouse keeper, William Darling knew he must try to save them. Grace insisted on accompanying him, and over her mother's objections, the two set out in a twenty-foot rowboat to try to save the survivors. On drawing close to Harker's Rock, they realized that there were more people than they had anticipated, a total of eight men and one woman who was clutching the lifeless bodies of her two children. Grace struggled against the waves to steady the boat while her father helped the woman and four of the men into the craft. After returning to the lighthouse and dropping off his charges, including Grace, William Darling and two of the crewmen he had just rescued rowed back to the rock and retrieved the other four men. In all, a total of eighteen people survived the wreck, five passengers and thirteen of the crew. The exact number of lives lost was never known. There were originally twenty-two crew members and around forty passengers. A passenger list, if one existed, did not survive the wreck. The best estimates held that about forty-three lives were lost.

Although Grace Darling's actual participation in the rescue was limited to her first spotting the wreck of the ship and assisting in one of two trips to gather survivors, her story began to take on a life of its own.

Within days word spread of survivors being saved. Local news reporters were quickly on the scene. Amongst them was a journalist from the *Berwickshire and Kelso Warder* who met with some of those rescued by the Darlings. One survivor wept as he disclosed that a young woman in a rowing boat had come to save them. The story immediately made the local newspapers. News of this act of outstanding courage spread like wildfire throughout the country. Within days *The Times* had published an account. The papers described in detail the violence of the storm and the tragic fate of the *Forfarshire* and its passengers, but how against all the odds Grace, with her father, had been able to save nine souls from the wreck. It fired the imagination of the public who declared her a heroine. One correspondent to *The Times* wrote: "Is there in the whole field of history, or of fiction even, one instance of female heroism to compare for one moment with this?" In addition, the point was made that she had risked her life for people she did not know—for strangers.[17]

Over the ensuing months and years, Grace's reputation continued to grow and spread nationally and internationally. She was eagerly sought out by artists interesting in doing her portrait or the scene of the rescue of the *Forfarshire* survivors. Songs were written about her, as were poems, including one by William Wordsworth, whose brother was one of those lost in the wreck. Sight unseen, she received offers of marriage as well as invitations to appear on stage and in at least one circus. The frenzy of her popularity became so intense that the Duke of Northumberland, with Grace's father's permission, became her guardian in order to protect her from her throngs of admirers. Newly crowned Queen Victoria sent Grace fifty pounds in honor of her heroism. The Royal Humane Society awarded her a gold medal. Over the decades, at least twenty books have been written that include Grace Darling as a character. The 2018 novel *The Lighthouse Keeper's Daughter*, for example, is a historical romance that connects the story of Grace Darling's rescue of the *Forfarshire* with a twentieth-century descendant of one of the passengers she helped rescue. A *Publisher's Weekly* online review describes it as appealing "to fans of low-key women's fiction."[18]

The notoriety took its toll on Grace. She was uncomfortable with the attention, preferring the solitude of her island home and lighthouse lodgings. In the early 1840s, her health began to decline, ending in her death on October 20, 1842, of tuberculosis. By that

Fig. 7D "The Wreck of the Forfarshire" by Thomas Musgrave Joy (1840). A figure representing Grace Darling can be seen in the boat.

time, however, her name had become synonymous worldwide with the heroic actions of a young girl intent on saving the lives of others. In 1876, for example, sixteen-year-old Grace Bussell assisted in the rescue of fifty people from a ship that had foundered off the coast of western Australia, earning her the sobriquet of "Grace Darling of the West." In 1882, Roberta Boyd, the twenty-year-old daughter of a lighthouse keeper in New Brunswick, Canada, rescued two men whose sailboat had overturned in the St. Croix River. The Canadian Department of Marine and Fisheries presented Boyd with a new boat, describing her as the "Grace Darling of the St. Croix." America, too, had Ida Lewis, its own Grace Darling, whose rescues and exploits would exceed those who preceded her.

Idawalley Zoradia Lewis, known to all as simply "Ida," was born in Newport, Rhode Island, in 1842. In 1857, she moved with her family to the Lime Rock lighthouse in Newport Harbor, where her father, Captain Hosea Lewis, was the keeper. After only a few months there, Captain Lewis suffered a stroke, leaving him disabled. With her mother, Ida took up the duties of tending the light, acting as substitute keepers for their impaired husband and father. When Captain Lewis died in 1873, Ida's mother was appointed official keeper, and Ida stayed on to assist as before. Ida was appointed keeper of the Lime Rock light in 1879 following the death of her mother the year before. She would continue to serve in that position until her death in 1911.

In all, Ida Lewis spent some fifty-four years of her life first as the daughter of lighthouse keepers, then as a keeper herself. During this time she was officially credited with saving eighteen lives, although some estimates put the figure as high as twenty-five. Over the decades, she became the nineteenth-century equivalent of a rockstar, with her exploits repeated time and again in news and magazine articles for nearly half a century. Frequently described as "America's Grace Darling," Lewis allegedly made her first rescue when she was scarcely a teenager—varying accounts have her as being between twelve and fifteen years of age. By her late twenties, she'd achieved sufficient fame to warrant having her image on the front of *Harper's Weekly*, accompanied by an article praising her courage. (See Figure 7E.) In November 1881, an engraving depicting one of Lewis's dramatic rescues juxtaposed next to

Fig. 7E Ida Lewis in "Harper's Weekly," July 31, 1869.

domestic scenes of her attending the lighthouse lamp filled a full page of *Frank Leslie's Illustrated Weekly*. (See Figure 7F.) With time, the accounts of her various rescues took on an almost apocryphal quality, frequently marveling at the fact that a mere *woman* could perform such deeds. *Harper's* described one rescue as "a daring feat, and requiring courage and perseverance such as few of the male sex are even possessed of."¹⁹ A *Leslie's* reporter opined "that Miss Lewis

Fig. 7F Ida Lewis in "Frank Leslie's Illustrated Newspaper," November 5, 1881.

has also developed an independence of courage as shown by her deeds, which prove also that the isolation of her life has not in any way prevented the development of tenderness of sympathy with suffering which is supposed to be peculiar to only the helplessness of women."[20]

The Lime Rock lighthouse was located well within Newport Bay, a short trip by boat to the shore. As Lewis's fame spread, visitors by the thousands flocked to the lighthouse to meet her, often bringing gifts to show their admiration. During the summer of 1879, she told a correspondent for the *Baltimore Sun* that "five thousand people perhaps have taken the trouble to come from shore to visit me and the season is not over yet by any means." With that, she showed him gifts she had received, "a number of articles supposed to appeal to the feminine heart—silks, scarfs, hose, gloves, bonnets, dresses and even jewelry," opining that "I probably shall never wear one quarter of them."[21] In addition to personal gifts there were numerous public accolades. To name a few of many, the secretary of the Treasury awarded her a gold medal for saving two soldiers from drowning; the citizens of Newport presented her with a boat named *Rescue*; and the Massachusetts Humane Society and the New York Lifesaving Benevolent Association voted to honor her with medals. In acknowledgement of her fame, Lewis's boat, medals, and other personal memorabilia were proudly displayed in the Rhode Island pavilion at the 1893 Chicago World's Fair.

In an early example of celebrity endorsement, what appeared to be a news article on Lewis appeared in the November 5, 1893, edition of the *Daily Inter Ocean*, a Chicago newspaper. Headlined "America's Grace Darling, the Brave Woman Who Keeps the Lime Rock Light," the account began as a description of Lewis and her lighthouse home. In contrast to other news accounts that had focused on her strength and stamina, Lewis was quoted as saying, "I have never been well in my life until now. My trouble was in my chest and lungs, and I have always had a cough from a child. What has done me more good than anything else in the world is Paine's Celery Compound." Lewis continued, "I have always been miserable in summer, and I believe I should have died this season if I had not taken Paine's Celery Compound. I began with it last February and this summer I have been splendid." The article went on to say how Lewis had lost faith in doctors' prescriptions and other patent medicines, but on taking Paine's Celery Compound, "I received benefit from the very first bottle." The article *cum* advertisement concluded by stating, "Ida Lewis is called the bravest woman in America. But there are many brave women who have suffered as she has done, who are nearly broken down, who need today a true food for the brain and nerves, and whom Paine's Celery Compound will make well again. Thousands of other women have been saved by this wonderful remedy."

As with lighthouses, designed and built to function as navigational beacons, the roles played by lighthouse keepers have frequently taken a life of their own, morphing into other realms of symbolism, notoriety, and meaning.

PART TWO

LIGHTHOUSES OF THE GEORGIA COAST

CHAPTER 8

TYBEE LIGHTHOUSE

The Tybee lighthouse, together with its southern sister, St. Simons light, is one of the crown jewels of the Georgia coast. Not only is it notable for its long and fascinating history, it is also one of the more complete light stations still in existence along America's Atlantic shores. It was one of the original lights ceded to the new United States government after the Revolution. Its history during the turbulent Civil War era is important, and it still serves as a landmark and lighted beacon marking the entrance to the Savannah River estuary. There are excellent on-site exhibits as well as the nearby Tybee Island Museum and Fort Pulaski National Monument, of which the Cockspur lighthouse is a part.

Tybee Light traces its history to the first half of the eighteenth century. The current lighthouse is the fourth in a series of navigational towers that have stood at or near its current location since the 1730s. Tybee Island was a natural choice for a landmark to help guide the way to the new colony of Georgia, established in 1733. The easternmost of Georgia's barrier islands and rising on average some seven and a half feet above mean sea level, Tybee's northeast corner might have been the first sight of land a ship would see when approaching the Savannah River from the east. In early 1733, shortly after arriving from England with the first group of colonists, James Oglethorpe chose this site to construct a wooden beacon[1] tower as a lookout and landmark. Although he ordered the construction to be

---

[1] Although the word "beacon" in current usage commonly implies the presence of a light, in older terminology it often referred to a prominent landmark or signal tower, as was the case of the first tower built on Tybee.

Fig. 8A The Tybee Light Station as maintained by the Tybee Island Historical Society.

completed as soon as possible, on his return from a trip to England three years later, it still was not finished. Following Oglethorpe's threats to the chief carpenter's life, "within sixteen days, more work was done than had been done in the preceding sixteen months," in the words of one historian.[1] The tower was finished a few weeks later in the spring of 1736. Built of pinewood on a brick foundation and octagonal in shape, the structure stood ninety feet high with a base width of twenty-five feet, tapering to twelve and a half feet at its apex. The lower twenty-six feet were covered with weatherboarding, with an open framework above.

At the start of the construction of the first beacon tower, Oglethorpe settled ten families on Tybee Island. Isolated from the main colonial settlement seventeen miles upriver and surrounded by marshland in the midst of a breeding ground of malarial mosquitos, by the fall of 1737 the island was deserted. While the tower remained, there was no one there to attend to its maintenance. It fell into increasingly worse disrepair. By 1741, there were serious discussions regarding repairs, all of which came to naught as the tower collapsed during a storm in August of that year. Almost immediately work began on a second tower. Built in the same area as the first, it stood 94 feet high, with a 30-foot-high flagstaff protruding from the top, for a total height of 124 feet. The entire structure was covered with weatherboarding. Its completion was celebrated with speeches and beer on March 24, 1742.

Although more sturdily constructed than its predecessor, this new Tybee beacon suffered from its location. The constant seaborne winds took their toll, as did shifting sands and beachfront erosion from periodic storms. By 1758, the sea was "lapping at the very foundations" of the tower, having progressed some thirty feet inland in ten years.[2] During this time the head river pilot had been given charge of maintaining the beacon tower, a near impossible task as there was no housing for him or his family on Tybee Island. In 1758, a retaining wall to protect the beacon tower and a house for the pilot were built, but the eventual outcome seemed inevitable. The recurring lack of available funds for repair and renovation made matters worse.

In 1768, the Provincial Assembly voted to raise funds for a new and sturdier lighthouse on Tybee Island to be located further from the edge of the sea. This time the tower would be made of brick rather than wood. Completed in early 1773, it was octagonal in shape and approximately a hundred feet in height on a twenty-four-foot-wide base. It is clear that the first wooden tower built on Tybee was unlighted and served only as a landmark. Whether the second wooden tower was lighted at times is uncertain. The new brick tower appears to have been designed to function as a traditional lighthouse but apparently remained unlit until ownership was transferred to the United States government in 1791.

On December 15, 1791, the Georgia legislature approved and Governor Edward Telfair signed an act to cede ownership of the Tybee lighthouse and five adjoining acres to the United States. This transfer of ownership came with provisions, however. The federal government was required to keep the light "in proper repair," supply "the necessary lights," and continue to provide funding of "three pence per ton for clearing and removing wrecks and other obstructions" in the Savannah River. Now under the aegis of the Treasury Department, a keeper named Higgins was appointed, maintenance and improvements were assured, and the lighthouse would now be integrated into a national system of navigational aids. George Putnam, in his 1917 *Lighthouses and Lightships of the United States*, noted, "Spermaceti candles were used in the lantern, because on account of its small diameter the smoke obscured the light when oil was used; this appears to be the only record of the employment of candles for lighthouse illumination in this country."[3]

By 1838, the lighthouse was tersely described as "a fixed light, fifteen lamps, fifteen-inch reflectors. Height of lantern above the sea, one hundred feet. Height of tower from base to lantern, ninety-five feet."[4] In 1841, the light was retrofitted with sixteen-inch reflectors. With the transfer of American lighthouses to the newly created Lighthouse Board in 1852, Tybee and most other lighthouses were refitted over the next few years with the vastly superior Fresnel lens system. Tybee's turn came in 1857 when the oil lamp and reflector system were removed and a second-order lens installed. For the next four years, the now-powerful light would guide mariners safely into Tybee Roads, but across a divided America, storm clouds of war were on the horizon.

The Confederate Lighthouse Bureau was created by the Provisional Congress of the Confederate States of America on March 6, 1861. As noted in Chapter 6, in contrast to the United States Lighthouse Board on which it was based, the CSA Lighthouse Bureau's primary initial directives involved dismantling the system of navigational lights along the coasts of the Confederacy in response to anticipated military action on the part of the North. To this end, the bureau contracted with Linville & Smedberg, a Savannah firm of "Machinists and Engineers," to remove the second-order Fresnel lens that had been installed in the Tybee lighthouse in 1857. A June 1861 document from the National Archives records the transaction with an invoice from the firm to the "Confederate States Light House Department" for the sum of $267.17 for "removing lights & boxing & packing" them. (See Figure 8B.) H. H. Linville, apparently one of the principals in Linville & Smedberg, billed ten dollars per day for ten and a half days for his time, in addition to the assistance of "4 extra workmen and machinists," whose work was charged at five dollars per man per day. Packing of the light required one bale of cotton (539 pounds at 3 cents per pound), 300 feet of one-inch

Fig. 8B An invoice sent to the "Confederate States Light House Department" by a Savannah engineering firm contracted to remove the lens apparatus from the Tybee lighthouse.

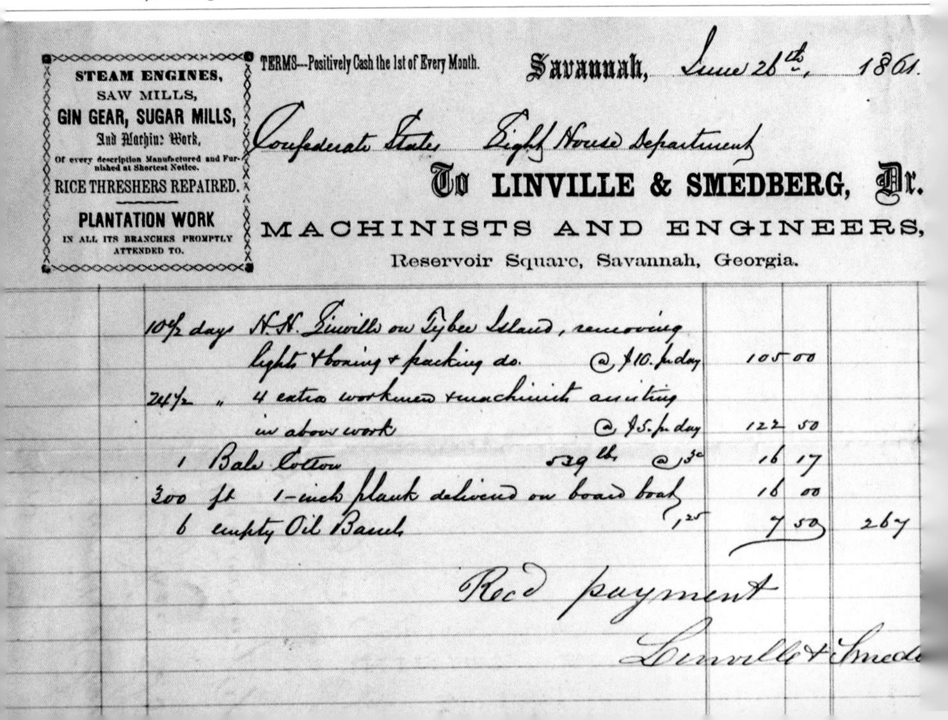

STEAM ENGINES,
SAW MILLS,
GIN GEAR, SUGAR MILLS,
And Machine Work,
Of every description Manufactured and Furnished at Shortest Notice.
RICE THRESHERS REPAIRED.
PLANTATION WORK
IN ALL ITS BRANCHES PROMPTLY ATTENDED TO.

TERMS—Positively Cash the 1st of Every Month.

Savannah, June 26th, 1861.

Confederate States Light House Department

To LINVILLE & SMEDBERG, Dr.

MACHINISTS AND ENGINEERS,

Reservoir Square, Savannah, Georgia.

10½ days H. H. Linville on Tybee Island, removing lights & boxing & packing do. @ $10. pr day	105 00		
24½ " 4 extra workmen & machinists assisting in above work @ $5. pr day	122 50		
1 Bale Cotton 539 lbs @ 3c	16 17		
300 ft 1-inch plank delivered on board boat	16 00		
6 empty Oil Barrels 1.25	7 50	267	

Rec'd payment

Linville & Smed

thick planking, and "6 empty Oil Barrels." The fate of the light and related parts after their removal is unknown.

One of the more interesting stories involving Tybee Light during the Civil War recounts the covert raid by Confederate troops for the purpose of destroying or disabling the lighthouse tower to prevent it from being used by occupying Northern troops. Even though the light had been removed, the lighthouse could serve as a valuable observation post, especially of nearby Fort Pulaski. Port Royal, South Carolina, fell to the combined forces of the United States Army and Navy on November 7, 1861. Less than thirty miles north of Tybee Island, this major port would be useful to the North as a staging area for an attack on Savannah. Northern troops landed on Tybee Island on November 25, 1861. Colonel Charles H. Olmstead, the commanding officer of Fort Pulaski and of Confederate troops stationed on Tybee Island, recounted in his memoirs the following events that took place shortly thereafter:

The loss of Port Royal convinced the Confederate Authorities of the uselessness of attempting to hold an isolated Island like Tybee with the force at their command, against such a naval force as would probably soon be sent against it. Accordingly it was determined to evacuate the Post and the danger seemed so pressing that the withdrawal was made with something like precipitancy, the heavy guns not being removed or made useless in any way. After waiting a few days and seeing no signs of an advance of the enemy an expedition was sent down from the fort [Fort Pulaski], the guns were dismantled, loaded on a barge and successfully brought up as an addition to our own armament. Not very long after this was done two or three vessels appeared off Tybee Point convoying transports loaded with troops, some of whom we could see with our glasses disembarked upon the Island. I was anxious that the tall lighthouse should not be used by the Yankees as a point of observation, also that a house that stood at Lazaretto Creek, on the Western end of the Island, should not serve as a blind for operations against the Fort.[11]

---

[11] Olmstead is referring to a house on the banks of Lazaretto Creek, a large tidal waterway almost directly across the southern channel of the Savannah River from Fort Pulaski. It would have been useful to Union troops as an observation post.

Fig. 8C The burning of the Tybee lighthouse in November 1861. The building on the left is reputed to be a Confederate barracks. It was later modified for use as a cottage for the second assistant keeper. The small building to the right is a summer kitchen originally constructed in 1812.

That night therefore I sent Captain J. B. Read of the Irish Volunteers with a squad of his men over to destroy both of these buildings by fire. He did the work faithfully and well; after he had been gone about an hour we saw flames bursting from the summit of the lighthouse and its narrow windows. At once the gun boats opened fire and began shelling the woods, causing us considerable uneasiness for the safety of the gallant Captain and his men, but ere many minutes had passed the King house at Lazaretto began to burn also and in a short while after the little party returned, muddy, smoky and very tired, but safe. This expedition called for considerable nerve on the part of Captain Read as he could not tell at what moment he might find himself in the very middle of the enemy. Had he been discovered nothing could have saved his party from capture or death. We were disappointed about the lighthouse however, for although the fire entirely destroyed all the wood work of the interior the solid brick shaft was left standing like a chimney and in two weeks or so the enemy had rebuilt the stairway and established a signal station at the top.[5]

As will be noted in the following chapter, Fort Pulaski fell to Union forces on April 11, 1862.

Fig. 8D Architectural drawing showing the configuration of the newly rebuilt Tybee lighthouse in 1867. The portion of the 1773 tower removed in the reconstruction is noted in red crosshatching on the drawing on the left.

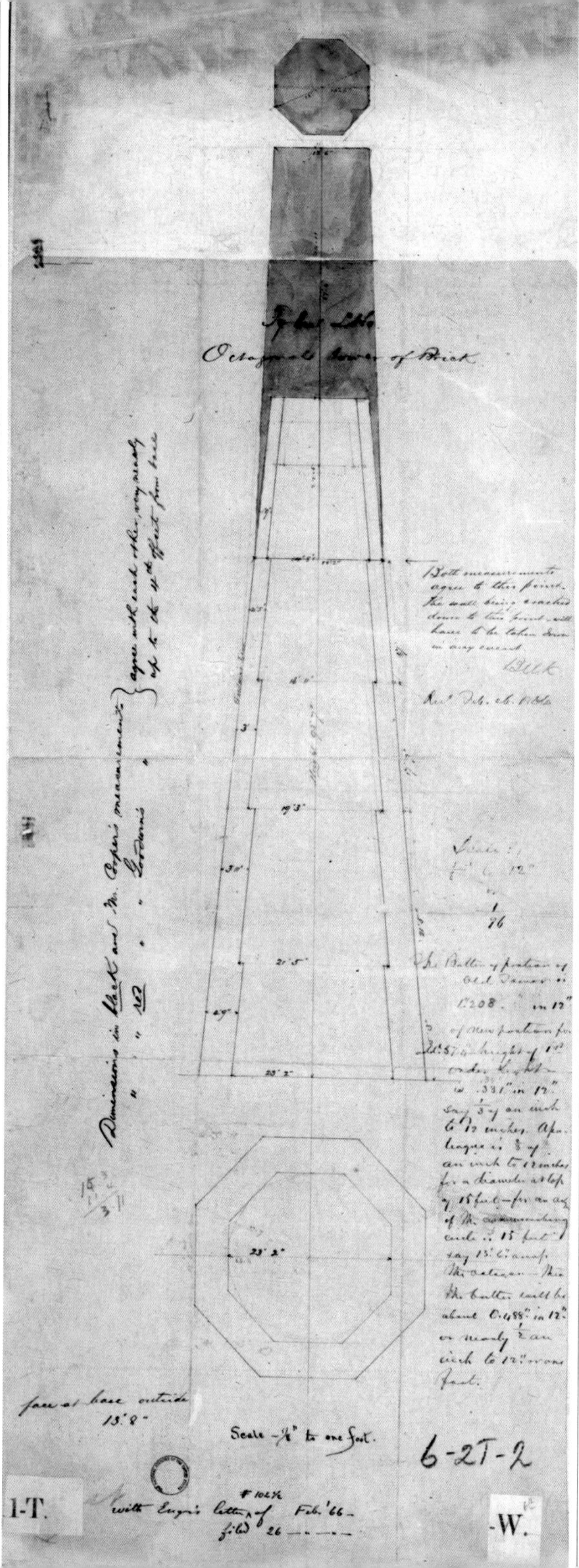

Octagonal Tower of Brick

Dimensions in black and Mr. Pope's measurements agree with each other very nearly up to the 10th offset from base.

Both measurements agree to this point. The wall being cracked down to this point will have to be taken down in any event.

D.C.K.
Recd Feb. 26, 1866

Scale
1" = 12'
1/16

The Batter of portion of old tower is 1.208" in 12"
of new portion for 57' in height of 1st order is ft is .331" in 12" say 1/3 of an inch to 12 inches. Average is 2/3 of an inch to 12 inches for a diameter at top of 15 feet — for an avg of the surmounting wall is 15 feet 6" and say 15' 6" an avg the between this, the batter will be about 0.488" in 12" or nearly ½ an inch to 12" or one foot.

face at base outside 15' 8"

Scale ⅛" to one foot.

1-T.

with Engr's letter of Feb. '66
filed 26

6-2T-2

-W-

Following the end of the conflict, appropriations totaling more than fifty-four thousand dollars were made in 1866 and 1867 for the rebuilding of the Tybee light and the construction of a nearby keeper's dwelling. While the Confederates had destroyed the wooden stairway and other vulnerable portions of the tower, its basic brick structure remained intact. After due consideration of the options, a decision was reached to use the existing 1773 structure as the base for a new and taller tower. The top portion of the older tower, which had stood approximately one hundred feet in height, was torn back to the point where new masonry construction could begin, raising the focal plane of the new light to 150 feet above sea level. Construction began in 1866 on the lighthouse and a new keeper's dwelling but was halted by an outbreak of cholera among the troops stationed on Tybee Island, leading to the death of the construction superintendent, H. D. Cooper, and four mechanics. Work eventually resumed with most of the progress in rebuilding the tower taking place the following year. A new first-order Fresnel lens was installed, and the light was put back into service on October 1, 1867.[6]

The newly refurbished and taller tower served its purpose well as a lighthouse, but one whose foundation was built on century-old brickwork. During the decade of the 1870s recommendations for a new structure were made annually to the Lighthouse Board. Over the following years, Tybee light survived a series of major storms in 1871, 1885, and 1893 as well as a number of less powerful blows. It was reported that "during the September 1878 gale, the tower vibrated to an alarming extent and the cracks, which had been pointed up, opened and extended."[7] Repairs were made to the light station's associated structures, but the tower itself, despite concerns about its age and fissured brickwork, survived unscathed.

A new head keeper's house was built in 1881, the same year that the switch was made from "lard oil" to "mineral oil" (kerosene) as the fuel for the light. Four years later a first-assistant keeper's

Fig. 8E (Opposite) A drawing of the plans for rebuilding Tybee Light in 1866–1867 made by H. D. Cooper, the construction superintendent, prior to his death from cholera in 1866. The original 1773 tower is seen at the base, with the portion to be added noted in blue.

dwelling was built, allowing individual housing for the head keeper and two assistant keepers.

The biggest challenge to the aging tower's integrity occurred on the evening of August 31, 1886. A massive earthquake centered just inland from the city of Charleston, South Carolina, shook much of eastern North America, followed by aftershocks that recurred for weeks. Estimated to be as high as 7.3 on the modern Richter scale, tremors were felt up and down America's East Coast and as far inland as New Orleans, Arkansas, and Ohio. Five hundred and fifty miles to the north, near Atlantic City, New Jersey, a lighthouse keeper found himself at the top of his tower when the earthquake struck. Unable to move because of wild swaying that was worse than "the fiercest gale of winter that ever struck the tower," he watched in horror as "hundreds and hundreds of plover, yellow-legged snipe, millet, English sparrows and strange sea birds circled wildly about the great lamp and dashed themselves in terror and in a way that I never saw them do before against the thick glass panes around the lamp until their blood dyed the glass."[8] In Savannah, only seventy-five miles to the south, people flooded into the streets and squares, fearing for their lives if they dared stay indoors. The tremors were apparently felt more intensely on Georgia's barrier islands, perhaps because of their alluvial makeup. In the Tybee lighthouse, "the lantern lenses…were broken and the machinery of the lamp was disarranged."[9] Preexisting cracks were extended, but the overall structure remained intact.

The end of the nineteenth century saw the construction of Fort Screven, a series of defensive batteries on the north end of Tybee Island. Begun in 1897 and completed the following year, the fort would eventually occupy 265 acres surrounding the lighthouse. At its peak the installation was armed with six batteries of coastal artillery designed to protect the entrance to the Savannah estuary. Like the lighthouse, however, the fort's nineteenth-century approach to defense would soon be made obsolete by advances in technology. The batteries were partially disarmed at the end of World War I and declared surplus in 1944 near the end of World War II. One of its former batteries now houses a museum with exhibits of local history.

Fig. 8F The Tybee lighthouse in 1885. It was painted white until 1887.

Fig. 8G The Tybee lighthouse in August 1914, shortly after its paint scheme was revised.

The advances of the twentieth century slowly arrived at Tybee light. In 1916, electricity was supplied to the head and first assistant keepers' cottages. The second assistant keeper's cottage would be electrified shortly thereafter, though this position was eliminated in the early 1920s. With the Great Depression and the fiscal austerity of the early 1930s, there was discussion of discontinuing the Tybee light station all together, or, as an alternative, supplying electrical power and automating the light, eliminating the need for full-time keepers. In April 1933, the position of the first assistant keeper was discontinued, and the following month the tower's beacon was converted to electricity. George Jackson became the head keeper and would serve until his death in 1947, when operation of the lighthouse was taken over by Coast Guard personnel. The beacon was fully automated in 1972 and as such remains as an aid to navigation.

Though the Tybee lighthouse continued to serve as a prominent seamark by day and illuminated beacon by night, its utility as a navigational aid became less important, supplanted by other more accurate means of location finding. In 1987, the Coast Guard vacated the Tybee light station, moving its local headquarters to

Fig. 8H The View of the Tybee Light Station from the top of the lighthouse tower. Within the white picket fence the head keeper's cottage can be seen near the top of this photo. To the left and right, respectively, are the first and second assistant keeper's cottages, with the small summer kitchen just beyond the former. The red-roofed structure to the lower left is the fuel storage building dating from 1890. Ticket sales and entry are via the white-roofed structure to the left, adjacent to the parking lot. In the distance, two of the former artillery batteries of Fort Screven can be seen.

nearby Cockspur Island, the site of Fort Pulaski. The Tybee Island Historical Society and the City of Tybee Island took over the light station's premises under a lease from the Coast Guard. In 2002, under the auspices of the National Historic Lighthouse Preservation Act of 2000, Tybee light station was among the first group of six lighthouses to be transferred from the federal government to other entities because of their cultural, recreational, educational, and historic value.[iii] The Tybee Island Historical Society assumed ownership and the responsibility for conserving and maintaining this important historic treasure.

It is impossible to fully appreciate the Tybee light station without visiting it. It is a place where history comes alive and allows one to imagine the life of a lighthouse keeper in the late nineteenth and early twentieth centuries. The lighthouse itself is a hybrid of the original brick 1773 tower topped by "new" construction in 1866–1867. The buildings that surround it, cottages for the head and two assistant keepers, have been beautifully restored and maintained. The summer kitchen adjacent to the head keeper's cottage dates from 1812.

The most prominent structure is the lighthouse tower, 145 feet tall and accessible via a spiral staircase of 178 steps. From the top, the view extends for miles in all directions, and it is possible to examine the massive first-order Fresnel lens from a close perspective. The lens, some nine feet in height and six feet in diameter, is made up of 320 glass prisms. Beyond its size, perhaps the most visual aspect of the tower is its paint scheme. The upper and lower thirds are black with the middle third painted white. This pattern helps identify Tybee light during daylight hours (i.e, as a daymark). The original post-Civil War tower of 1867 was painted white. In an effort to give Tybee light an unmistakable identity, in 1887 the bottom portion of the tower was painted black. In 1914, a black band was added just below the tower's lightroom and extended downward in 1917 such that the upper and lower thirds were black with the center third painted white. In 1965, the pattern was changed once again, with the upper two-thirds of the tower painted gray and the lower one-third white. The gray was overpainted with black in 1970 and changed again in 1999, when the paint scheme reverted to the 1917 pattern. A visual display of this quite confusing sequence of patterns can be found just outside the lighthouse entrance. (See Figure 8I.)

INFORMATION FOR VISITORS

The Tybee Light Station is easily accessible and visitor-friendly. From Savannah, follow the signs to Tybee Island via US 80 eastward, which becomes the Islands Expressway. You will pass Fort Pulaski National Monument on the left, which provides access to excellent viewing of the Cockspur lighthouse. Approximately 2.9 miles past the entrance to the fort, exit to the left on Polk Street, cross over Solomon Avenue, and turn right on Fort Avenue. Proceed straight ahead across Van Horne Avenue, at which time the street name changes to Taylor Avenue. The Light Station complex will be just ahead on the right. Parking is provided at no charge.

Tickets to visit the Light Station may be purchased in a white building on the edge of the parking lot. This former garage has been converted to serve as an entry point and gift shop.

Included in the ticket price is entrance to the Tybee Island Museum just across the street in one of the batteries of the former Fort Screven. Visiting hours are from 9:00 A.M. to 5:30 P.M. daily except Tuesdays. (The last tickets are sold at 4:30 P.M.) For current information and a list of admission prices, consult the Tybee Light Station website: www.tybeelighthouse.org.

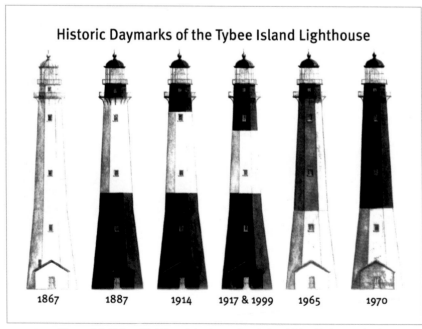

Fig. 81 Since 1867 Tybee Light's paint scheme has helped identify it as a daytime landmark ("daymark") for mariners. The current pattern, first used in 1917 and readopted again in 1999, has changed several times over the years.

CHAPTER 9

COCKSPUR LIGHTHOUSE

In selecting a site for the first city of the new colony of Georgia, James Oglethorpe bypassed the low-lying coastal barrier islands and chose instead a plot of high ground known as Yamacraw Bluff further inland on the Savannah River. It was there that the wharves and port facilities were established, becoming with time the center of commerce and trade for the region. To reach the city, however, ships needed to travel seventeen miles up one of two channels of the broad but potentially hazardous river. Tybee Light served to guide mariners into the broad estuary of the Savannah, but almost immediately the waterway divided into north and south channels, the former broader but more tortuous, the latter narrower but more direct. Separating the two channels today are three large islands, Cockspur Island nearest the ocean, then Bird Island and Elba Island, whose northern side borders a sharp ninety-degree turn in the course of the stream.[I]

Cockspur Island, perhaps so named for its shape, divides the river and in the process guards the entrance to both channels. Its location made it a frequent point of rest for travelers arriving from Europe. John Wesley, the evangelist who would later found American Methodism, spent two weeks on Cockspur[II] in 1736, delivering his first American sermon there. Additionally, the island

---

[I] As might be expected, river islands tend to change their number, shape, and configuration with time based on current, water volume, etc. Since the eighteenth century, Cockspur Island has changed perhaps the least.

[II] Then known as Peeper Island, possibly so named for the sounds made by native tree frogs.

Fig. 9A The Cockspur lighthouse in June 1885 showing the dock, which allowed access during high tides.

was a logical location for defense of the river as well as governmental control of smuggling and immigration. In 1761, a palisaded hundred-foot-square defensive position named Fort George was constructed around a thirty-foot-high blockhouse, only to be abandoned in the next decade.

British attacks during the War of 1812 again emphasized the need for coastal defense. Although a secure defensive fort in the area was originally authorized in 1816, twelve years passed before surveys were made and an exact location chosen. Among those assigned the initial engineering work was a young lieutenant and recent West Point graduate named Robert E. Lee. Lee oversaw the initial construction of the new fort's infrastructure before being transferred to other posts in 1831. Construction of Cockspur Island's massive fortress, to be named Fort Pulaski for the Revolutionary War hero and Polish patriot Count Casimir Pulaski, began in 1833 and was finally completed in 1847. The structure's thirty-two-foot-high walls varied in thickness from seven to eleven feet, requiring approximately twenty-five million bricks, most of which were made locally. With the fort completed, it was believed that entrance to the Savannah River could easily be denied to an enemy in time of war.

But it was not military vessels that plied the Savannah in the first half of the nineteenth century. It was sailing and later steam ships that ferried America's major export of cotton to the ports of Europe and elsewhere, returning with expensive consumer goods for the newly wealthy planter class and others whom the rising economy had enriched. To aid navigation on the river, in 1826 Congress voted an appropriation for a navigational beacon to be constructed "on Grass Island off the tip of Cockspur." In 1834 and 1837, additional appropriations were made for the construction of two beacons on the island. During the next two years one brick tower was built on Cockspur, but shortly thereafter damaged in a storm.[1]

In the late 1840s, one of the beacons on Cockspur Island was converted to a lighthouse under the direction of John S. Norris, a New York architect who designed and built a number of prominent structures in Savannah including the Savannah Custom House (1848–1852); the Andrew Low House (1849), associated with the foundation of the Girl Scouts of America; and the Mercer House (1859–1866), the scene of the murder recounted in *Midnight in the Garden of Good and Evil*. Officially established in 1849, the beam of the lighthouse was located twenty-five feet above sea level, generated by five lamps with fourteen-inch reflectors. It was said to be visible from nine miles away. This lighthouse was destroyed by the hurricane of 1854.

The current Cockspur Island lighthouse was constructed on the foundation of Norris's 1849 light. Built primarily under the supervision of the newly established Lighthouse Board, six thousand dollars was set aside to construct the new and taller tower as well as a keeper's dwelling nearby on Cockspur. Based this time on plans supplied by the board, Savannah Gray bricks were used as the primary construction material, with the ironwork cupula and lightroom shipped from Baltimore. It was completed and placed into service in 1856, presumably equipped with the then-new Fresnel lens system. The new tower was approximately forty-

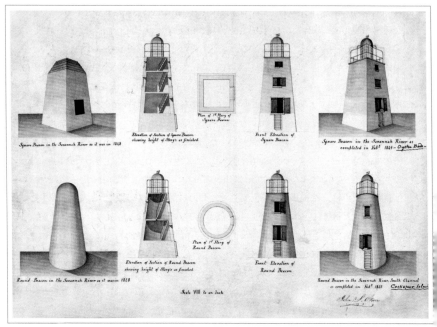

Fig. 9B Drawings from the National Archives of several beacons, Cockspur Island, in 1852. John Norris's lighthouse, destroyed in 1854, is seen at the lower right.

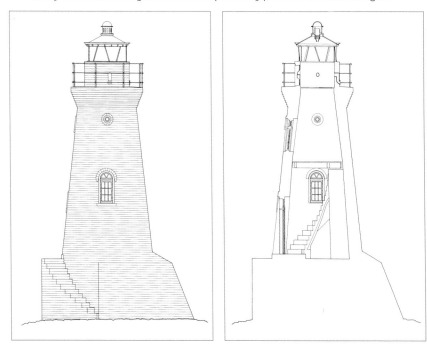

Fig. 9C Architectural renderings of the Cockspur light showing the base and external spiral staircase and **(Fig. 9D)** a cutaway view of the interior.

three feet in total height, with the focal plane of the light several feet below this level. As the lighthouse is located in an area that is flooded at high tide, the base level is approximately nine feet above grade and accessed via an external spiral staircase. The interior of the tapering tower has three levels. A brick spiral staircase ascends from the first to the second. The third, the location of the lightroom, is accessed via a wooden ladder and floor hatch. A narrow open gallery with rails surrounds the lightroom on the apex. The tenure of the new Cockspur lighthouse as a navigational aid was brief. The secession fever that had been simmering in the South was brought to a fore by Abraham Lincoln's election as president in November 1860. Even though Georgia's secession convention had not yet met, Governor Joseph Brown, fearing a possible attempt by the North to take control of the city and port of Savannah, ordered the seizure of Fort Pulaski.[2] This was accomplished without bloodshed on January 3, 1861, by members of the Savannah Volunteer Guards, the Oglethorpe Light Infantry, and the Chatham Artillery. With the surrender of Charleston's Fort Sumter in April 1861 and the institution of Union general Winfield Scott's "Anaconda Plan," the strategic importance of Fort Pulaski, as well as Georgia's coastal lighthouses, become all the more vital.[III] The Cockspur light was extinguished about the same time, along with most of the others along the coast of the Confederacy, as they might assist Northern forces while possibly illuminating Southern blockade-runners, who preferred the cover of night. The lighthouse would stay dark until after the end of the war.

Perhaps the most interesting fact about the Cockspur lighthouse is the fact that it even exists today. The city of Savannah was a plum prize for the North during the Civil War, and efforts to seize it began early in the conflict. As noted in the previous chapter, the Confederates had withdrawn defensive forces on Tybee Island at the entrance to the Savannah River, as the area was considered indefensible. Fort Pulaski had now become the first line of resistance. In late November 1861, more than twelve thousand Federal troops landed on Tybee with the goal of eventually taking the fort. General Robert E. Lee, then commander of Confederate forces in South Carolina, Georgia, and Florida, was familiar with Pulaski

---

[III] See Chapter 6, "Lighthouses during the American Civil War."

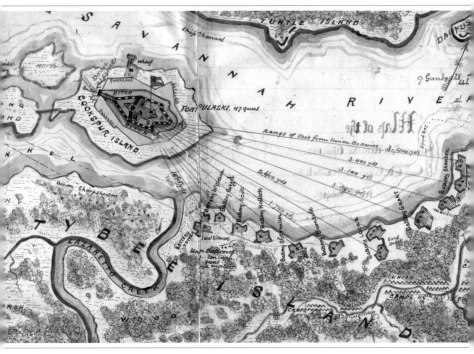

Fig. 9E A contemporary map shows the location of the artillery batteries on the north shore of Tybee Island in April 1862. Note that the Cockspur light, located at the southeast corner of the island, is directly in the line of fire of several of the batteries.

from his work there three decades earlier. He was confident that the garrison of 348 soldiers, forty-eight cannon, and six months of rations could withstand both an assault and a siege. The fort's walls were considered nearly impregnable to smooth-bore cannon bombardment, and its moat and open field of fire around it would be devastating to troops attempting to attack by land. Pulaski was left under the command of twenty-five-year-old Colonel Charles Olmstead. What Lee and Olmstead had not considered, however, was the use of a new type of weapon, the rifled cannon. The attack on Fort Pulaski would mark the first significant use of rifled cannon in combat.

During the following month, December 1861, Union naval forces began a blockade of the Savannah River. At same time, General William T. Sherman, commander of the forces occupying Tybee Island, set in motion a plan to take Fort Pulaski. As

Fig. 9F A photo showing the damage to one of the walls of Fort Pulaski from Union canon fire.

the winter and early spring of 1862 progressed, the opposing sides exchanged desultory fire, but no major battles took place. During this time, Union forces, working primarily under cover of darkness and camouflage, set up a total of eleven artillery batteries along the north shore of Tybee Island facing the fort. (See Figure 9E.) To the surprise of the Confederate troops, the first light of the morning of April 10th revealed a massive array of Union firepower aimed squarely at their supposedly impregnable redoubt. Shortly thereafter, under a flag of truce, a small boat appeared from the Union side carrying an emissary who demanded the surrender of the fort. Olmstead's reply was succinct and to the point: "Sir, I acknowledge receipt of your communication demanding the unconditional surrender of Fort Pulaski. In reply, I can only say that I am here to defend this Fort, not surrender it."[3]

The Union barrage began at 8:10 A.M. Over the next thirty hours some 5,275 rounds would be fired at the fort. As can be seen from Figure 9E, the Cockspur lighthouse was directly in the line of fire between the fort and six of the artillery batteries. After eighteen hours of bombardment, one the walls of the fort was breached, allowing explosive shells to land inside the garrison. As the shells began to fall near the north powder magazine and its forty thousand pounds of powder, Olmstead made the decision to surrender, striking the Confederate banner and running up a white flag at 2:30 p.m. on April 11th. The damage to the fort was massive, as can be seen in a contemporary photograph in Figure 9F.

Miraculously, the Cockspur lighthouse survived the battle essentially unscathed. To reach each other's positions, the Confederate and Union cannoneers were required to fire their weapons at an angle that fortuitously arched over the forty-three-foot elevation of the tower. Despite the lack of damage, it would remain dark for the remainder of the war. In April 1866, after the installation of a sixth-order Fresnel lens and some minor repairs that included painting the tower white, the Cockspur light's beacon once again marked the way to the Savannah River's southern channel.

Over the following decades, the lighthouse was damaged several times by the forces of nature. An August 1881 hurricane was accompanied by a storm surge that rose twenty-three feet above mean sea level, inundating part of the lighthouse tower, flooding Fort Pulaski, and destroying the lighthouse keeper's cottage nearby. Other major storms in September 1893 and October 1894 brought additional damage to the light. By the first decade of the twentieth century, the pattern of traffic on the Savannah River had changed. Once, both the north and south channels were in common use, but with time, the north channel became the preferred route to the upstream ports. In 1909, the Lighthouse Board's district engineer recommended discontinuing the tower's use as lighthouse, converting it instead to a navigational daymark. On June 1st of that year, after more than half a century in service, Cockspur Island's light was extinguished for the last time. The lighthouse remained the responsibility of the Lighthouse Bureau and subsequently the Coast Guard until 1959, when it was formally transferred to the National Park Service, where it became a part of the Fort Pulaski National Monument.

Fig. 9G View of the Cockspur lighthouse at high tide, ca. 2019.

Even though its light no longer shone, many of the stories, legends, and anecdotes associated with it lived on. One tale linked to the Cockspur light—albeit somewhat tangentially—is that of Florence Martus, Savannah's famed "Waving Girl." For more than forty years, day and night, she was said to have waved at every vessel that passed her home on the Savannah River, saluting them with a white cloth by day and a lantern in darkness. Martus was born on Cockspur Island in 1868, the daughter of a German immigrant named John Martus who was serving in the US military as an ordnance sergeant at Fort Pulaski. Her brother, George W. Martus, joined the Lighthouse Establishment at age sixteen in 1877. The census of 1880 lists his occupation as "assistant lighthouse keeper." In September 1881, he was named keeper of the Cockspur light, a position he held until the summer of 1884 when he was transferred to Elba Island, some half a dozen miles up the river nearer Savannah's port. There he was placed in charge of maintaining several

Fig. 9H Statue of Florence Martus, Savannah's famed "Waving Girl," located on the city's Riverfront promenade.

navigational lights along the heavily traveled maritime passage. In late 1884, Sgt. John Martus was transferred to Fort Omaha in Nebraska, at which time Florence, then aged sixteen, and her mother moved in with George Martus in the keeper's cottage on Elba Island. John Martus died in 1886. His wife and daughter Florence would live with George for the remainder of their lives.

Over the decades, a host of legends and apocryphal tales grew up around Florence Martus. She became variously known as the Lighthouse Girl, the River Queen, but most frequently as the Waving Girl. Between 1887 and 1931 she was said to have waved at more than fifty thousand ships and boats, in the process gaining international fame among members of the seafaring community. An oft-repeated romantic tale—one denied by Martus herself—was that of a lost love, a sailor who had stolen her heart and disappeared with an unkept promise to return one day. Martus's own explanation is even more bizarre.

Apparently, shortly after her move to Elba Island, Martus was stricken with diphtheria. According to a 1904 account in the *Savannah Morning News*,

The attack deprived her of her speech and hearing. For three or four years she was without these powers, and it was by a miracle that they were restored to her. In the year 1886 came the terrible earthquake that wrought such havoc and which was felt in considerable violence in Savannah. People here were terrified and so was the small lonely family at the lighthouse station. Mrs. Martus and her son and daughter feared while the earth beneath them was swaying and rocking that their last hour had come, but what seemed a calamity was really a blessing to them. The girl, moved by the shock of terror, instantly recovered her speech and hearing, and from that day to this she has retained them. It was during the period of her great affliction that her love for the passing ships was born, and through all the years since it has endured.[4]

In later interviews, Martus would date the beginning of the Waving Girl to the year 1887. Her obsession with never missing the chance to signal a passing vessel day or night was aided in part by her dogs. As told in one of many admiring news columns, they had "become imbued with the spirt of Elba Light, for they bark joyously when they hear the lapping of the waves that [a] ship causes."[5]

Beginning in the 1890s and continuing long after her death in 1943, dozens of newspaper articles were written about the Waving Girl, some of which clearly slipped into the realm of fantasy. In 1912, for example, a Charleston newspaper reported that she had helped save "crew after crew of vessels wrecked off Tybee shoals near Fort Pulaski" during an 1879 hurricane.[6] Martus would have been only eleven years old at the time. She was said to have "saved three men from drowning" during the 1893 hurricane.[7] In 1911, she assisted in saving eight men from a burning dredge in the Savannah River, for which she was presented a "gold-lined loving cup."[8] After the latter rescue, the *Baltimore Sun* noted, "When you honor the memory of Ida Lewis, keeper of the Lime Rock light, the 'Grace Darling of America,' do not forget Florence Martus of the Savannah River, for she is a heroine, too."[9]

When her brother George reached the mandatory retirement age of seventy in 1931, he and Florence gave up their Elba Island home and moved to a quiet cottage on the road to the Isle of Hope near Savannah. In February 1936, the famed columnist

Ernie Pyle sought her out after hearing the legend of the Waving Girl. He found George and Florence, then sixty-seven years old, sitting around a wood stove discussing the cold weather. Referring to Florence, Pyle recounted,

I'd half expected to see a mystic, unreal sort of person, legend-like even in personality. But she's quite matter-of-fact. She is very small and weathered. Her white hair is cut short like a man's, only in front it's longer and curled up across her forehead. She had on a brown sweater. "Is it true that you never missed waving at a single ship in forty years?" I asked her.

"Yes, I guess it is. Even longer than that. How long were we there, George?"

"Forty-four years," brother George said, sucking at his pipe. Brother George was the lighthouse keeper, and sister Florence kept house for him. Just the two of them, on Elba Island, in the Savannah River, for forty-four years.[10]

In 1938, the City of Savannah and Chatham County sponsored a birthday party in honor of Martus's seventieth birthday. More than three thousand people attended. She died in a Savannah hospital in February 1943. Later that year, a Liberty Ship built in Savannah by the Southeastern Shipbuilding Corporation was christened the *S.S. Florence Martus*. In 1971, a statue of Martus and one of her faithful collies sculpted by Felix de Weldon was erected on Savannah's Riverfront.[IV]

INFORMATION FOR VISITORS

Cockspur Lighthouse can and should be combined with a visit to the larger Tybee Lighthouse, located only two miles to the east on Tybee Island. US Highway 80, known in the city of Savannah as Victory Drive, changes names to the Islands Expressway as it heads east to its terminus on the southern tip of Tybee Island. Cockspur Lighthouse is today part of Fort Pulaski National Monument, accessed via a clearly marked entrance to the left (north) on the

---

[IV] Felix de Weldon is perhaps best known for his iconic statue of the US flag being raised on Iwo Jima, located at the US Marines Corps War Memorial in Arlington County, Virginia.

route to Tybee. The National Park Service collects an entrance fee from visitors. Information and exhibits pertaining to the lighthouse can be found on the grounds of the fort and in the visitors center near the parking area.

The lighthouse itself is not open to the public. The Lighthouse Overlook Trail, a well-marked walking path approximately one mile in length, runs from the parking area near the fort to the spit of land on which the lighthouse is located. At high tide, the terminal part of the trail is under water; at low tide it is possible to get within a few hundred feet of the tower, but under any circumstance, that section of the trail is quite muddy. The lighthouse can, of course, be accessed by boat, but entry is not allowed.

For a superb view of the lighthouse from the perspective of one of the batteries that bombarded Fort Pulaski in April 1862, there is a viewing area nearer Tybee Island. To reach it, drive 1.8 miles past the entrance to Fort Pulaski. You will see a small street on the left (north) named Battery Drive. Just inside the entrance there is an informal parking spot to the left and a viewing area with several informational signs erected by the National Park Service.

Fig. 91 Cockspur Lighthouse just after sunset, as viewed from Battery Drive.

CHAPTER 10

SAPELO LIGHTHOUSE

Sapelo, the fourth largest of Georgia's coastal barrier islands, lies some forty miles south-southwest of Tybee Island. Just over ten and a half miles in overall length and encompassing approximately 16,500 acres, human habitation of its sandy pine- and live-oak-covered forests dates back at least four millennia. During Spain's conquest of the New World, it was the home of Spanish missionaries as early as the sixteenth and seventeenth centuries and later formed part of the ill-defined contested border between England's American colonies and Spanish Florida. Formally ceded to the Province of Georgia by the Creek Indians in 1757, over the following decades a number of owners exploited the island for its timber and agricultural potential.

The name perhaps most prominently associated with Sapelo is that of Thomas Spalding, an innovative planter, politician, and financier who, beginning in 1802, would eventually acquire ownership of essentially the entire island along with extensive other holdings on the Georgia coast. Spalding raised sea island cotton, sugar cane, rice, and cattle, in the process becoming a wealthy man prior to his death in 1851. In 1912, Howard Coffin, then vice president of the Hudson Motorcar Company of Detroit, acquired most of the island from Spalding's descendants. Facing financial ruin in the early years of the Great Depression, Coffin sold his interest in 1934 to Richard J. Reynolds, the tobacco heir. In the years following Reynolds's death in 1960, the State of Georgia acquired ownership

Fig. 10A Opposite, the 1820 Sapelo lighthouse viewed from the east across the marsh separating the lighthouse site from the main body of Sapelo Island.

of the vast majority of Sapelo from his heirs. Today, the northern part is managed as a wildlife refuge, with much of the southern part of the island and its marshes established as a National Estuarine Sanctuary and home to the University of Georgia's Marine Institute.

The presence of a lighthouse on Sapelo Island is closely related to the economic history of Georgia in the years between Eli Whitney's invention of the cotton gin in the 1790s and the crash of the cotton economy and the Great Depression during the first third of the twentieth century. In colonial days, Georgia's first settlements were along the coast, later moving inland along the courses of major rivers, including first the Savannah, then the Ogeechee, and later the Altamaha. The Altamaha River is formed by the confluence of two other major rivers, the Oconee and the Ocmulgee, which meet eighty-five miles inland near present-day Hazlehurst, Georgia. With a floodplain covering approximately fourteen thousand square miles, the Altamaha Basin drains nearly one quarter of Georgia's territory. It is the third largest riverine source of fresh water to the Atlantic along America's east coast. Navigable to the shoals that mark the state's geologic Fall Line, until the early twentieth century commercial transportation was common as far upriver as Milledgeville on the Oconee[1] and Macon on the Ocmulgee.

As settlement and commerce spread across Georgia, transportation of goods to market relied on waterborne routes. As recounted in the application to list the Sapelo light on the National Register of Historic Places,

Starting in the late eighteenth century and continuing into the early twentieth century, the small city of Darien was a major port on the coast of Georgia, in some ways rivaling Savannah to the north and Brunswick to the south. Located at the mouth of the Altamaha River, the largest river on Georgia's coastal plain, ten miles inland from the coastal islands, the port of Darien served as a transshipping point for enormous quantities of agricultural produce, mostly rice and cotton, along with vast amounts of timber and naval stores from the vast south Georgia forests. During the late nineteenth century, the sawmills and lumber yards in and around Darien were among the largest in the world; yellow pine from Georgia was shipped from Darien to ports up and down the eastern seaboard and literally around the globe.[1]

---

[1] Milledgeville was Georgia's capital from 1804 until 1868.

Fig. 10B A late nineteenth-century view of the Sapelo lighthouse and keeper's cottage.

In May 1808, the Georgia Legislature ceded jurisdiction of five acres of land at the southern tip of Sapelo Island to the United States government for the purpose of building a lighthouse. In July 1816, the US Treasury Department acquired five acres from Thomas Spalding on which to build the new light. The location, on Doboy Sound, was significant in that it marked the entry to the preferred water route to the growing seaport of Darien. In September 1819, the Lighthouse Establishment signed an agreement with Winslow Lewis for the construction of a brick tower sixty-five feet in height to be topped by a fifteen-foot-high, windowed lantern room made of iron, as well as a keeper's dwelling near its base. The contract called for the tower to be round in shape, tapering from a diameter of twenty-five feet at its base to twelve feet at its apex. An addendum a few months later specified that the lamp room was

Fig. 10C A nineteenth-century photo of the Wolf Island rear-range beacon and keeper's house. The light was decommissioned in 1899. Its site has since been washed away by the sea.

to be fitted with an array of fifteen of Lewis's patent lamps, each with sixteen-inch reflectors, all mounted on a rotating triangular frame. The keeper's house was to measure twenty by thirty feet in size, divided into two rooms, each of which was to have a fireplace. The Sapelo light was first illuminated in early 1820.

Concomitantly, the Lighthouse Service acquired land and signed a contract to build two beacon lights on Wolf Island directly to the south across Doboy Sound from the Sapelo light. The smaller was a wooden tower while the larger, a small whitewashed tower thirty-eight feet in height, was made of brick and adjacent to a newly constructed keeper's cottage. Both were put in service in 1822, serving with the main Sapelo lighthouse as navigational aids for ships traversing the dangerous shoals near the mouth of the sound.

For the first four decades of its existence, the Sapelo light, aided by the lights on Wolf Island, guided sailors in and out of Doboy Sound. In 1854, the Lighthouse Board replaced Lewis's inefficient lighting system with a new fourth-order Fresnel lens broadcasting a light that flashed every forty-five seconds. Around the same time, the tower was painted in a distinctive pattern of red and white alternating bands allowing it to serve as an unmistakable daymark. To further assist navigation, an additional range light, initially constructed of wood, was erected to the southeast of the lighthouse. With the start of hostilities of the Civil War, the Sapelo light, like most other lighthouses along the coasts of the Confederacy, was disabled, its lighting apparatus removed and hidden in late 1861 or early 1862.[11] At the end of hostilities, the Lighthouse Board made extensive repairs to the lighthouse and keeper's cottage, replacing the lens system and reigniting the light in April 1868. In 1890, a brick oil house was constructed adjacent to the base of the lighthouse to hold fuel for the light's lanterns.

Fig. 10D 1877 cast-iron range light.

The wooden range light tower, constructed in the 1850s, continued to require significant maintenance. As a result, the decision was made in 1877 to replace it with a square pyramidal cast-iron tower located 660 feet distant from the lighthouse. This beacon still exists and is said to be unique in that the enclosed square lantern room is the only known range beacon of this type to exist in the United States today.[2]

The greatest threat to the Sapelo light arrived in the form of the hurricane of 1898. At the time, it was the strongest such storm in recorded history to have struck the Georgia coast. With sustained winds of 135 miles per hour, the hurricane made landfall on Cumberland Island on October 2nd. Accompanied by a massive storm surge, the confirmed death toll included at least 179 victims.

---

11 The Wolf Island beacons were disabled during the early days of the Confederacy and were said to have been "blown up" in 1863. As with most other lighthouses darkened during that era, they were rebuilt and relit in the postwar period.

Fig. 10E The 1820 Sapelo lighthouse, oil house, and cistern in 1996, prior to their restoration.

Brunswick, the largest coastal population center hit by the storm, saw floodwaters rise sixteen feet. At the Sapelo lighthouse, pounded by thirty-five-foot waves, the keeper and his family took refuge in the top of the lighthouse tower while water rose inside to a level of eighteen feet.[3] At the Wolf Island beacons less than two miles to the south across Doboy Sound, the island was said to be "completely swept and several lives reported lost."[4] Moving rapidly inland, the storm continued to release massive amounts of rain on Georgia, Florida, and the western Carolinas. Highlands, North Carolina, for example, reported a storm-related deluge of 12.5 inches.

As the storm cleared, the severity of the damage to Sapelo light was evident. The keeper's dwelling was uninhabitable. The wind-driven waves and storm surge had undermined the tower's

Fig. 10F The 1820 Sapelo lighthouse after its 1998 restoration.

Fig. 10G An aerial photo of Sapelo's 1905 steel-frame lighthouse (top) and the remains of the 1820 brick tower lighthouse (bottom).

foundation. A post-storm survey by the Lighthouse Board's district inspector indicated that maintaining the Sapelo lighthouse in its current location would require significant repairs, with the alternative being the construction of a new lighthouse at a less threatened location. On the south side of Doboy Sound at Wolf Island, the erosion caused by the storm necessitated rebuilding and moving the beacon, but the degree of damage led to the decommissioning of this light in early 1899.[III]

In 1902, Congress appropriated forty thousand dollars for the purpose of constructing a new lighthouse and keeper's dwelling in the vicinity of the 1820 lighthouse. By 1904, work had begun on a new hundred-foot steel skeleton lighthouse and a pair of keepers' dwellings to be located approximately 750 feet to the northeast of the brick tower. In September 1905, the transition was made, with

---

[III] The receding coastline has erased the former site of the Wolf Island light.

the beacon of the old lighthouse extinguished and that of the new tower illuminated for the first time. The old keeper's cottage had been dismantled, leaving at the old site only the brick tower, the oil house, and a cistern formerly used to collect water for the keepers.

Much of the traffic in Doboy Sound during the nineteenth and early twentieth centuries was generated by the shipment of cotton, and later by products of the timber and naval-stores industries. In the years that followed World War I, cotton production fell sharply due to the boll weevil, even as railroads supplanted waterborne transport of the remainder. Around the same era, timber harvesting began a precipitous decline due to depletion of forest resources and lack of demand resulting from a slumping national economy. By 1930, Doboy Sound, once bustling with maritime traffic, saw little commercial shipping as business shifted to the more accessible ports of Savannah and Brunswick. In 1933, the Lighthouse Service made the decision to decommission Sapelo light, effective in June of that year. The following year, the hundred-foot steel tower was dismantled and shipped to Lake Michigan, where it was re-erected as the South Fox Island Light.[IV] The two keepers' dwellings were dismantled and sold as scrap, leaving the site to be marked only by a brick oil house and the concrete footings of the tower.

For most of the remainder of the twentieth century, the location of the former Sapelo lighthouses remained abandoned. The 1820 tower fell into disrepair, shedding its brightly colored red-and-white-banded stucco to reveal its underlying brick structure. Spurred by an interest in preserving this unique historical monument, a movement began in the 1990s to restore the lighthouse to its configuration of a century earlier. Application was made and granted to include the site on the National Register of Historic Places. Funded by both public funds and private donations, the lighthouse and adjacent oil house were restored over an eight-month period in 1998. The tower received a fresh coat of stucco and red and white paint and new steps were fabricated for its interior.

[IV] The South Fox Island light was automated in 1958 and remained in service until 1969 when it was decommissioned. Although nonfunctional, the tower still stands and is owned by the state of Michigan.

Fig. 10H Interior of the 1820 lighthouse tower prior to its restoration in 1998. The spiral cypress steps, each individually cut to fit the tapering structure, were badly deteriorated.

The oil house was reroofed and secured against the elements. Other components of the former light station, including the 1877 range light and the water cistern that formerly served the keeper's house, were preserved. Explanatory signage describing the lighthouse's purpose and history were erected. At a ceremony on September 6, 1998, Governor Zell Miller inaugurated the restored light, turning a switch that relit this historic beacon for the first time in nearly one hundred years.

INFORMATION FOR VISITORS

Sapelo Island is a unique place, in many ways unspoiled by the modernity that has overwhelmed many beachfront islands. There are no roads or bridges connecting the island to the mainland; all access is via scheduled ferry service operated by the state of Georgia. Although it is possible to rent a private dwelling on the island, most casual visits are made via scheduled tours conducted year-round under the auspices of Sapelo Island National Estua-

Fig. 10I The restored 1820 Sapelo lighthouse.

rine Research Reserve and operated by the Georgia Department of Natural Resources and the National Oceanic and Atmospheric Administration. Lighthouse tours are conducted each Saturday. These tours normally include guided walks to the top of the brick tower as well as the former site of the early-twentieth-century steel tower that was dismantled in 1933. From a historical perspective, Sapelo lighthouse is important for the fact it is one of the few remaining lighthouses constructed by Winslow Lewis, and the oldest among those of his that have survived to the present day.

The Sapelo ferry departs from the mainland dock adjacent to the Sapelo Island Visitors Center at 1766 Landing Road in Darien. There are three round trips per day Monday through Saturday and two round trips on Sunday. The ferry does not run on Georgia state holidays. Reservations must be made in advance. The center and dock are located just off Georgia Highway 99, approximately ten miles east of Interstate 95's exit 58. For information and to book individual tours, the visitor center's number is 912-437-3224. Group tours and field trips on Sapelo can be booked by calling the Sapelo Reserve directly at 912-485-2300.

While access to the interior of the lighthouse is available only as part of a guided tour, the lighthouse grounds are open and can be freely visited. Though considered part of Sapelo Island, the site itself was on a large hammock[V] originally separated from the main portion of the island by marsh. For a number of years motor vehicle access to a parking area for lighthouse visitors has been available via a gravel road off the causeway that connects the Reynolds mansion to Nanny Goat Beach. In addition to the lighthouse, oil house, cistern, and 1877 range light, there are remnants of an Endicott Era artillery battery erected around the time of the Spanish-American War to protect Doboy Sound.[VI] Beginning near the base of the 1820 lighthouse, a path leads to the site of the 1905 steel-frame lighthouse site. Here the remains of an oil house that served that light can be seen, and an observation tower has been erected to give the visitor an overview of the surrounding area.[VII]

V "Hammock" is a term most commonly used in Southeastern American English to describe a small, tree-covered islet, commonly surrounded by or adjacent to marsh or swampland. It likely derives from "hummock," a sixteenth-century seafaring term used to describe a small protrusion of land (e.g., hill or hillock) seen over the horizon.

VI Between 1885 and 1905, a series of coastal fortifications were erected to protect certain American waterways from hostile naval action. Sites were chosen by a joint military board appointed in 1885 by President Grover Cleveland and headed by Secretary of War W. C. Endicott, leading to the name of these emplacements. Fort Screven near Tybee Lighthouse is another example of an Endicott Era fort.

VII Note that the initial restoration and repair of the Sapelo lighthouse was completed in 1998. During the more than two decades that have elapsed since, the elements have taken their toll on the lighthouse, educational signage, and access road from the main part of the island. At the time of the writing of this book in the spring of 2020, repair and repainting of the Sapelo lighthouse was underway and access to the interior of the lighthouse tower was not available. Also, it appears that the path and wooden walkway to the 1905 lighthouse site, some of which cross marshy areas, are being rebuilt. This situation should change within a matter of time. Prospective visitors should inquire as to the status of repairs and renovations prior to scheduling a visit to the site.

CHAPTER 11

ST. SIMONS LIGHTHOUSE

The St. Simons light station, as with the Tybee lighthouse complex, is located in a popular tourist destination and is readily accessible to visitors to the Georgia coast. Architecturally intact and magnificent in its detail, it is situated on a plot of land whose history dates to the earliest years of the Georgia Colony. The St. Simons Island of today bears little resemblance to that of the eighteenth century when this area of British North America lay on the disputed border with Spanish possessions in Ponce de Leon's *La Florida* to the south. The colony of Georgia was founded on the basis of philanthropic and altruistic ideals under a 1732 royal charter granted to the "Trustees for Establishing the Colony of Georgia in America" (whose number included James Oglethorpe) by King George II, for whom the colony was named. On a practical basis, however, the Crown and Parliament saw the establishment of settlements of "sturdy farmers" in this area as a barrier to the northward expansion of the territorial ambitions of Spain, Britain's rival in the New World.

With Oglethorpe as their leader, the first group of colonists arrived in America in January 1733. By mid-February, Oglethorpe had chosen an inland site on a high bluff facing the Savannah River for what would become the colony's first city, its name taken from the river.[1] As military defense of the colony was a primary concern,

---

[1] The name "Savannah" for the river, and subsequently the city, is said to be derived from a variant on the name of a Native American tribe that settled an area near present-day Augusta. Alternatively, it has been suggested that the name derived from the English word "savanna," or marshy grassland, which itself was adopted from *sabana*, a Spanish word of similar meaning.

Oglethorpe built a series of defensive forts, most notably Fort Frederica on St. Simons Island, some sixty miles south of Savannah. The construction of the fort and an adjacent town of the same name was approved by the trustees in 1735 and completed over the next several years, representing at the time the southernmost point of British colonization in North America. The location of the fort on the Frederica River bordering the landward side of St. Simons Island was chosen for its advantages as a defensive installation protecting the southern approach to the Altamaha River.

While Fort Frederica was under construction, British soldiers under the command of Lt. Phillip Delegal constructed fortifications on the extreme southern tip of St. Simons Island.[11] This site guarded the approach to the Frederica River as well as St. Simons Sound and its inland tributaries. In 1738, Delegal's troops were absorbed into Oglethorpe's regiment and the original fort rebuilt and enlarged as Fort St. Simons, only to be destroyed by Spanish troops in 1742. It

Fig. 11A The St. Simons lighthouse and keeper's cottage.

11 Then known as Sea Point and later as Couper's (or Cooper's) Point.

was near this same site, nearly seven decades later, that St. Simons's first lighthouse would be built.

As early as 1738, British colonists were settling the land around St. Simons Sound. After the defeat of a Spanish military expedition by Oglethorpe's troops in the battles of Gulley Hole Creek and Bloody Marsh in July 1742, the threat of outside military intervention lessened. The following year Oglethorpe returned to Britain, leaving day-to-day management of the colony in the hands of the trustees, even as problems mounted. In 1752, frustrated by their seeming inability to fully implement the lofty goals on which the colony was founded, the trustees ceded ownership back to the British crown. On January 2, 1755, Georgia officially became a royal province.

With the 1763 Treaty of Paris that ended the French and Indian War,[III] Britain effectively gained control of North America east of the Mississippi River, including the Spanish colony of La Florida. Later that year, King George III issued a proclamation extending Georgia's southern border from the Altamaha to the St. Marys River, thus opening new lands for settlement. In 1771, the Province of Georgia purchased a thousand-acre tract on the East River just off St. Simons Sound in order to establish a planned city similar to Savannah. It was to be named Brunswick, honoring the Duchy of Brunswick-Lüneburg and the Germanic heritage of the Hanoverian kings, including George III.

Following the American Revolution, coastal Georgia saw a steady influx of settlers, drawn to the new lands for the cultivation of sea island cotton and rice as well as the exploitation of timber and other forest resources. Brunswick's growing importance in trade was recognized during the first session of the new American Congress in July 1789. An act establishing a series of customs collectors designated the city as one of Georgia's four official ports of entry for the new nation.[IV] An act of the Georgia legislature in

---

[III] The French and Indian War refers to the North American theater of the larger global conflict between European powers known as the Seven Years War.

[IV] In addition to Brunswick, ports of entry were also designated at Savannah, St. Marys, and Sunbury. The Act stated that "the duties and fees to be collected… shall be received in gold and silver coin only," setting the value of the "Mexican dollar" at 100 cents. As such, "Mexican dollars" (silver eight-real coins) remained in common circulation in the United States through the mid-nineteenth century.

1797 transferred the Glynn County seat from the town of Frederica, by then nearly abandoned, to Brunswick. Eli Whitney's invention of the cotton gin and the resulting explosive growth of the cotton culture in the early 1800s transformed Brunswick and other seacoast towns into major export centers with increasing maritime traffic in and out of potentially treacherous estuaries.

In recognition of the increasing importance of St. Simons Sound and the port of Brunswick, in May 1804 Congress enacted a law directing the secretary of the Treasury (under whose purview lighthouses fell in those days) to construct a light on the southern end of St. Simons Island. In October of that same year, John Couper sold to the federal government for the nominal sum of one dollar a four-acre tract on Couper's Point, the former location of Fort St. Simons. Some two and half years later, Congress authorized nineteen thousand dollars to build the lighthouse and adjacent keeper's dwelling. In early 1808, the contract to construct the lighthouse was awarded to James Gould, a native of Massachusetts. The specifications called for a tower seventy-five feet in height (excluding the lantern room), constructed of tabby,[V] and erected on a stone foundation. The tower, octagonal in shape, was to taper from a width of twenty-five feet at the base to ten feet at its apex. The exact nature of the lighting apparatus is unknown, but it can be assumed that it was similar to others of the day, a chandelier-type arrangement of lamps burning whale oil. The lighthouse was completed in 1810. Gould, justly proud of his work, applied for and was appointed as lighthouse keeper by President James Madison in May of that year, drawing a salary of four hundred dollars per year. He would serve in that position for the next twenty-seven years, retiring in 1837.[VI] In 1816, during Gould's tenure as keeper, Winslow Lewis

---

[V] "Tabby" is a building material composed of ground oyster shells, lime, and sand mixed with water to produce a concrete-like substance when dry. Tabby was widely used in coastal Georgia in the eighteenth century in lieu of other forms of masonry.

[VI] In addition to his duties as lighthouse keeper, James Gould became a landowner and cotton planter. A fictional character of the same name based on Gould is the major character in Eugenia Price's 1972 novel, *Lighthouse*, the first book of her St. Simons Trilogy (though published last).

replaced the tower's original oil lamps with a nine-light version of his patented system of modified Argand lamps and reflectors.

With the formation of the Lighthouse Board in 1852, a program was undertaken to refit all American lighthouses with the

Fig. 11B The 1810 lighthouse constructed by James Gould. It would be destroyed by Confederate forces in 1861.

vastly superior Fresnel lens system. In 1857, Winslow Lewis's lights were replaced by a French-made third-order Fresnel lens beaming a fixed white light. Unknown to all at the time, it would be in service for only four years.

With the winds of war in the air and on orders from Georgia governor Joe Brown, the Jackson Artillery departed Macon by train on January 24, 1861, assigned to guard the entrance to St. Simons Sound. Carrying with them their armament of four six-pound cannons, two twelve-pound howitzers, and "a full complement of Minnie muskets," they were escorted to the train depot by a military band while keeping step to the music of "The Girl I Left Behind Me," with "a large crowd of ladies and others accompanying."[1] After an overnight trip to Savannah, the regiment was billeted at one of the city's finest hotels and entertained at a banquet that evening by the Chatham Artillery. Departing by steamer to St. Simons Island the next morning, they were "greeted by cheers at every point" as their ship made its way to the open sea.[2] The regiment set up camp near the St. Simons lighthouse on Couper's Point, the location of Fort St. Simons more than a century prior. The artillery battery and earthworks, known as Fort Brown, were set up near the shore to guard the entrance to the sound. The lighthouse was taken over as a powder magazine on "demand of Col. Thos. Burke, Aide-de-Camp to the Governor,"[3] and the keeper's house was taken over as lodging for the officers of the artillery.[VII]

A letter written by a member of the company recounted his initial days on St. Simons as one might describe a holiday on the coast:

The residents of this beautiful little island, Messrs. King, Couper, Postell, Gould and others, met us when we disembarked Saturday night at Hamilton, and have shown us every attention in their power. They, with their carts, horses and servants, have been at our service ever since we landed. They have furnished us with wood for our camp-fires, shucks for our mattresses, loads of oysters, mullet and turnips for our tables, and

---

VII The status of the lighthouse beacon is unclear. The blockade of Southern coasts was not instituted until April 1861, this being the primary reason for the extinguishment of coastal lights. It is reasonable to assume that the keeper continued to maintain the light at this early date.

are now building saw palmetto tents to relieve our small canvas tents of a portion of their occupants. Some of them are in camp every day, devising something to aid or amuse us, and we find all of them agreeable and cultivated gentlemen.[4]

The horrific reality of the coming war was yet to be recognized.

The idyllic mood would soon change with the institution of the blockade of Southern coasts ordered by President Abraham Lincoln on April 19, 1861. The rebellious South, while wealthy due to the production of cotton, was at its heart an agrarian society. Unlike the increasingly industrialized North, the economy of the nascent Confederacy was dependent on the income from the exports of its products, which, in turn, would be used to purchase the matériel of war. A blockade would cut off this source of funds, weakening the ability of the South to defend itself or wage war. By late 1861 and early 1862, it became increasingly evident that the geography of the Georgia coast was not ideal for defensive positions, leading to the withdrawal of most troops and the abandonment of seaside fortifications. In order to deny aid to the enemy, lighthouses—which would serve as navigational beacons for the blockading Northern fleet—were ordered darkened, with many destroyed in order to prevent their being used as observation platforms. On September 29, 1861, St. Simons lighthouse keeper Isham W. Hawkins recorded in his log that "the light tower was blown down" by Confederate forces and the navigational buoys marking the approach to St. Simons Sound had been sunk. Sometime prior to this—the exact date is unknown—the third-order Fresnel lens installed in 1857 had been removed and stored on shore in Brunswick.

Shortly after sunrise on the morning of March 9, 1862, three gunships from the Northern blockading fleet crept carefully into St. Simons Sound, expecting at any moment to be fired upon by the Confederate batteries on Couper's Point adjacent to the St. Simons lighthouse, and those on Jekyll Island on the south side of the sound. To the evident surprise of Commander S. W. Godon,[VIII] the officer in charge, the batteries were silent. It was soon discovered that they had been abandoned and their artillery removed, possibly

---

VIII Some reports spell the commander's name as "Gordon."

several months earlier. A landing party was put ashore at Brunswick, only to find the town deserted, with the wharf and rail depot in flames, said to be the work of retreating Confederate soldiers. The buoys marking the navigational channel had been moved, and a search for missing lighthouse lenses, which Godon had been told were stored somewhere in the town, was fruitless. A few days later Godon's steamship, accompanied by another vessel, made its way north via "the inland passage" (presumably by the now-abandoned Fort Frederica) to the mouth of the Altamaha River and Doboy Sound. Darien, like Brunswick, was said to be deserted.[5] Coastal Georgia would remain under effective control of Federal forces for the remainder of the war.

The end of hostilities brought Georgia's coastal lighthouses, buoys, and other navigational aides back under the control of the Lighthouse Board. The task before them was a daunting one. To one degree or another, all of the lighthouses had been disabled, either by removal of their lights and lenses or by destruction of the towers themselves, as in the cases of St. Simons and Tybee. On February 10, 1868, the Treasury Department published a detailed proposal calling for bids to rebuild the St. Simons light station. Instead of the octagonal shape of the 1810 tower formed from tabby, its replacement was to be a tapering cone constructed of brick. The specifications called for a height of ninety-three feet, with diameters of approximately twenty-two feet at the tower's foundation and eleven feet at its apex. On top of this, a lantern room containing a third-order Fresnel lens would place the focal plane of the beacon one hundred feet above ground level. The proposal also specified a 2,400-square-foot keeper's dwelling, connected directly to the lighthouse tower via a passageway that served as a site for the storage of lamp oil. The entire complex was to be constructed about thirty feet north of the ruins of the 1810 lighthouse destroyed by the Confederates. (See Figures 11C, 11D, 11E.) The contract for construction of the lighthouse and keeper's dwelling was awarded to Charles B. Cluskey on November 20, 1868. Cluskey was a well-known architect and builder responsible for the design and construction of a number of notable buildings in antebellum Georgia, including the Old Medical College in Augusta (1837), the Governor's Mansion in Milledgeville (1837–1839), and St. Vincent's Academy in Savannah

ST. SIMONS LIGHTHOUSE

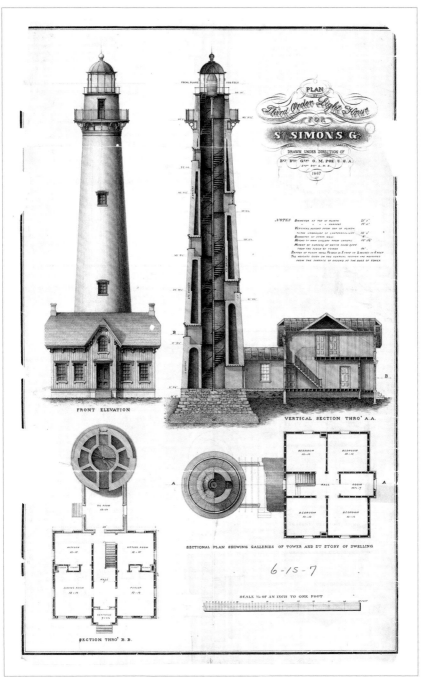

Fig. 11C The request for bids to construct a new lighthouse after the Civil War included detailed plans of the light tower and keeper's cottage.

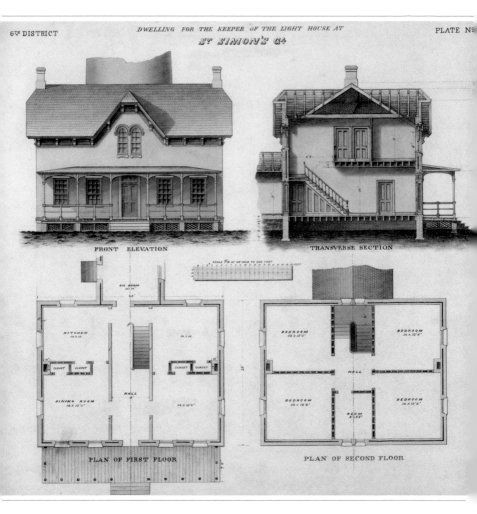

Fig. 11D Plans for the keeper's cottage from the 1868 request for construction bids.

(1845). Unfortunately, Cluskey succumbed to an acute illness in July 1871 when the tower was approximately half completed. His death was attributed to malaria, common at the time on the coastal isles.[IX] Though delayed, the project was completed and the beacon illuminated for the first time on September 1, 1872. In contrast to

---

IX Cluskey's bondsman-successor also died of an illness believed to be malaria, further delaying construction.

Fig. 11E The new St. Simons lighthouse and keeper's cottage under construction circa 1870–1871. Note that to the right of the photo, the stump of the 1810 lighthouse, destroyed by Confederate forces in 1861, can clearly be seen.

the fixed light of the earlier lighthouse, the new beacon alternately flashed red and white lights at one-minute intervals.

The new light station displayed a unique architectural beauty, especially the keeper's quarters. Later described as "a good example of mid-Victorian, Romantic Eclectic architecture," combining "a Gothic Revival roof with an Italianate bracketed cornice and paired gable windows, enhanced by Italianate drip arches," the building appears today essentially the same as it did in 1872.[6] Anecdotally, the cottage was reputed to be the only brick structure in Glynn County prior to 1880. The size and configuration of the finished dwelling would allow two families to occupy the structure under somewhat cramped circumstances, a situation that might have contributed to tragedy a few years later.

The end of the Reconstruction Era brought relative prosperity to coastal Georgia. In addition to cotton, the timber and naval-stores industry provided steady sources of employment, thriving local businesses, and healthy maritime trade. The role of the newly constructed St. Simons lighthouse as a guide for seaborne traffic was vital to the local economy. Bradford B. Brunt served as the keeper of the light from July 1872 until he was replaced by Frederick Osborne in April 1874. Osborne's assistant keeper was John Stevens, appointed to that position in March 1876. Osborne and his family occupied the ground floor of the keeper's quarters while Stevens and his family lived upstairs. Bad blood had developed between the men, the exact cause uncertain. Some reports suggest that Osborne had spoken inappropriately to Stevens's wife.[7] According to family members, however, the disagreement was over chickens. Both men kept a flock, which was said to have led to an argument when one accused the other of stealing his fowl.[8] Whatever the cause, at about 8:30 on Sunday morning, February 29, 1880, Stevens shot Osborne, fatally wounding him. According to the local weekly newspaper,

It seems that there had been bad feeling between these gentlemen for several days, and on Sunday morning they went out into the bushes in front of the house to settle their difficulty. During this interview Stevens threatened to chastise Osborne, when Osborne drew his pistol and ordered him not to advance further, whereupon Stevens went back into the house, took down his double barreled shot gun (which had been previously loaded with buckshot for deer hunting) and as Osborne advanced along the path near the fence, leading to the gate, Stevens fired, at a distance of ninety-eight feet, hitting him in four places, only one shot, however, taking serious effect.[9]

The account went on to remark that although Osborne was "suffering very much," he would "probably recover." The writer commented, "Of the merits of the case, or who is to blame in the transaction, we forbear [sic] to express ourself, as it will undergo judicial investigation."

The following week's edition of the newspaper carried two interesting paragraphs interspersed among a column of local news. The first noted "Frederick Osborne, who was shot in the Stevens-Osborne tragedy, an account of which we gave last week, died last Wednesday [March 10th] at 3:30 p.m." The second paragraph, seemingly not connected in any way to the first, read, "It is a fact worthy of much praise that as soon as he shot Osborne, Stevens went at once for medical assistance for the wounded man, repaired immediately to Brunswick and gave himself up to the proper authorities, and the very moment the bond for his appearance was signed he returned to his post as assistant light-keeper, and has been on double duty ever since, performing both his and Osborne's duty incessantly day and night."[10] Records from the National Archives indicate that John Stevens was "removed" from his position of assistant keeper on March 16, 1880.[x] A Brunswick jury, after reviewing the circumstances of Osborne's killing, was said to have acquitted him of a charge of murder.

The killing of Frederick Osborne seems to have helped spawn the legend that the St. Simons lighthouse is haunted. In an interview with a reporter for the *Atlanta Journal* in 1968, the widow of lighthouse keeper Carl Olaf Svendsen spoke of a ghostly presence that manifested itself in the form of footsteps clearly heard treading up and down the iron steps of the tower.[11] The ghost was heard but not seen, as the footsteps would always stop, never going beyond the top landing or the bottom of the stairway. Legend attributed the footfalls to "one of the early lighthouse keepers who returns to see that the light is being cared for properly." Referring to "the sound of the invisible footsteps" as "one of best authenticated of the island ghost stories," Mrs. Svendsen said her late husband, who served as keeper between 1907 and his death in 1935, would tell her that as he sat in the tower minding the light, "I thought you were coming up to visit me tonight, but when the footsteps came no higher than the top landing, I knew it wasn't you."

Another major memorable event of the 1880s was the Charleston earthquake of August 31, 1886. Isaac Peckham, the keeper at the time, reported that at 9:30 P.M. he was in the watch room of the tower lying down when the tower began to sway from northwest to southeast. The first shocks lasted two and a half minutes, breaking one of the flash panels in the light and accompanied by a noise "like that made by a horse running over a hard

---

[x] Osborne's replacement as head keeper was George W. Asbell, appointed on June 24, 1880, for a salary of $600 per year. Asbell resigned in 1883. In 1915, he was one of six men killed in Brunswick in a mass shooting by a "real estate and timber dealer," apparently precipitated by financial losses in a land transaction. (*New York Times*, March 7, 1915.)

Fig. 11F The St. Simons lighthouse and keeper's cottage in 1886. Note the absence of the oil house (constructed in 1890) and the distance from the current shoreline.

road."[12] Aftershocks occurred nightly for the next three weeks, but there appeared to be no significant damage to the integrity of the tower.

In the 1880s, most American lighthouses transitioned to "mineral oil" (kerosene) as the primary fuel for their lamps, the change bringing with it an increased risk of fire from this far more flammable fluid. In 1890, a new stand-alone brick oil house was constructed adjacent to the base of the lighthouse tower. Similar structures were built around the same time at Tybee and Sapelo lights. The lighthouse survived the great hurricane of October 1898 that caused massive flooding and a number of deaths in Brunswick and among the coastal islands. In September 1904, the *Brunswick Daily News* reported on its front page that the lighthouse had been struck by a bolt of lightning, destroying the telephone, breaking a number of windows, "and striking a fine Jersey milch cow belonging Keeper [Joseph] Champagne, which was killed instantly."[13, XI]

While still acting as an important beacon by night and landmark by day, St. Simons's significance as a navigational aid began to decline in the twentieth century as technological advances offered other, often more precise, methods of location finding. The beacon was switched to electricity in 1934 and fully automated sixteen years later, marking the retirement of the last keeper. Brunswick and St. Simons Sound remained an important seaport for the export of cotton, timber, and naval stores for many years, and during World War II, became a major site for the building of Liberty Ships.

Couper's Point, the former location of Fort St. Simons and the current location of the lighthouse and keeper's dwelling, has changed significantly over the years. The current shoreline is more than four hundred feet inland from that of a century ago, eroded away by the constantly changing flows of currents and tides and further sculpted by the occasional hurricanes that strike the coastal islands. The sites of the former fort and the dock that once served the lighthouse have been lost, their locations having been reclaimed by the sea. Fortunately, the lighthouse and its associated structures, located further inland, have been spared.

Today, the St. Simons lighthouse complex is managed by the Coastal Georgia Historical Society. A recent article in a journal devoted to lighthouses described it as "one of the prettiest, most complete, and best-interpreted lighthouses in the country." In 1972, the federal government deeded the keeper's dwelling to Glynn County for use as a museum and visitor's center, now managed by

---

XI "Milch" is a Middle English word properly used as an adjective to describe a domestic animal that yields milk, e.g., a milch goat or a milch cow. In contemporary common usage, however, a more commonly used term would be "milk cow."

Fig. 11G Left, the St. Simons lighthouse today. Visitors can climb the hundred-foot tower for a magnificent view of St. Simons Sound. In this photo, the third-order Fresnel lens can be seen in the lightroom. Right, **(Fig. 11H)** the plaque on the St. Simons lighthouse lens indicates it was manufactured in Paris by L. Sautter et Cie.

the historical society under a fifty-year lease. The light tower itself is owned by the society, acquired in 2004 through the National Historic Lighthouse Preservation Act. Over the recent decades, the lighthouse and keeper's cottage have undergone several renovations. The cottage now serves as a living museum, and visitors have easy access via a spiral stairway to the top of the hundred-foot tower. As it has for the last century and a half, the Fresnel lens beams out its beacon nightly, its lenses and lamps lovingly maintained by the local Coast Guard Auxiliary.

The Keeper's Dwelling Museum traces the development of lighthouse technology and helps visitors discover the fascinating history of coastal Georgia through colorful exhibits of rare artifacts, historical photographs, and interactives designed for all ages. The period rooms on the second floor of the dwelling are decorated as they might have been in the year 1907, when the lighthouse keeper and his family lived and worked here in the days before electricity and indoor plumbing.

The highlight of a visit to the St. Simons lighthouse is the opportunity of climbing the tower to the top for a view of St. Simons Sound from the balcony. While the lightroom atop the tower is not

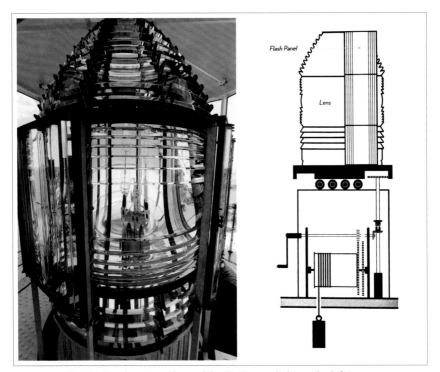

Fig. 11I The third-order Fresnel lens of the St. Simons light on the left is surrounded by a rotating frame (or flash panel) that focuses the beacon. Although powered today by electric motor, originally the frame would have been moved by a windup mechanism powered by weights, as shown in the illustration to the right.

open to the public, it is possible to get a closer look at the 1872 lens from below. The third-order Fresnel lens was manufactured in Paris by L. Sautter et Cie.[XII] (See Figure 11H.) The company was the successor to the firm of François Soleil, an instrument maker who worked closely with Augustin Fresnel in the early development of the new lens types. Soleil began commercially manufacturing lighthouse lenses in the 1820s. In the 1830s and 1840s, his firm was managed by relatives until it was sold in 1852 to Louis Sautter, who changed the name to Sautter et Cie. In 1870, Paul Lemonnier joined the firm as a partner, changing the name to Sautter, Lemonnier et Cie. The plaque on the St. Simons lens would suggest that it was

[XII] "Et Cie" is the French abbreviation for "et Compagnie," or, in English, "and Company."

manufactured in Paris between 1852 and 1870, at least two years before being first illuminated on the Georgia Coast.

Another item of interest is the mechanism that projects the lighthouse's beacon as a rotating light. The risk of mariners lost at sea mistaking one lighthouse for another is real and resulted in many shipwrecks during the eighteenth and nineteenth centuries. During daylight hours, a distinctive paint pattern served to identify individual lights, for example, the red and white bands of Sapelo or the black and while color pattern of Tybee. The St. Simons tower was white. During hours of darkness, the St. Simons light installed in the 1870s alternately displayed red and white flashes at sixty-second intervals. Today, only a white beacon is broadcast, but the rotating mechanism that produces periodic flashes remains intact. (See Figure 11I.)

INFORMATION FOR VISITORS

A visit to the St. Simons lighthouse, keeper's cottage, and museum is a must for visitors to St. Simons Island. The lighthouse complex is located just to the rear of the magnificent A. W. Jones Heritage Center, which serves as the headquarters of the Coastal Georgia Historical Society (CGHS). The heritage center is located at 610 Beachview Drive, approximately three hundred yards east of St. Simons Pier and Village. Parking for the building is available on both sides, but access to the lighthouse is more convenient from the Twelfth Street parking area on the building's left. Tickets for the lighthouse and museum can be purchased in the heritage center's gift shop, or online via the CGHS's website at coastalgeorgiahistory.org. Additional information is available on the website, or by calling (912) 634-7090. School and group tours are welcome with advance notice.

CHAPTER 12

LITTLE CUMBERLAND LIGHTHOUSE

The lighthouse on Little Cumberland Island is the least well known of Georgia's five existing coastal beacons, yet for more than seventy-five years it served to guide mariners into the treacherous waters of St. Andrew Sound,[1] surviving hurricanes, war, and the relentless pounding of the sea before being decommissioned in 1915. Privately owned and not accessible to the general public, its colorful and interesting history is often overlooked, though, as one historian noted, "the story of Little Cumberland lighthouse is, in large part, the story of nineteenth century life along this remote section of Georgia's southern coast."[1] Listed on the National Register of Historic Places, the lighthouse likely owes its very existence today to the preservation efforts of a group of dedicated individuals who have restored and maintained it over the last half century.

The Cumberland Island complex is the southernmost and largest of Georgia's barrier islands, measuring roughly 18.5 miles in length from north to south and 3.5 miles at its greatest width. At the northern end, Little Cumberland Island is separated from Cumberland Island proper by a tidal creek and salt marsh. Little Cumberland measures approximately 3.5 miles in length and slightly more than a mile in width. The south end of Cumberland Island, facing Amelia Island, Florida, marks the entrance to Cumberland Sound and the St. Marys River estuary. On the northern end, Little Cumberland Island forms the southern border of St. Andrew

---

[1] Also often referred to as St. Andrew's *or* St. Andrews Sound.

Fig. 12A The Little Cumberland lighthouse today as seen from St. Andrew Sound.

Sound, with Jekyll Island immediately to the north. St. Andrew Sound, in turn, leads to the mouth of the Satilla River and to Jekyll Sound, with tidal river access to St. Simons Sound.

In the mid-eighteenth century, this area of what is now modern Georgia lay on the disputed frontier between Spanish Florida and British Georgia. Even with the end of the French and Indian War in 1763 and the transfer of Florida to British control, the areas around the Satilla and St. Marys rivers remained wild frontier; settlement grew slowly. The end of the American Revolution brought settlers, but the treaty that ended the war also returned Florida to Spanish control. Life on the frontier could be dangerous. "Criminal fugitives and runaway slaves traversed the area to escape in Florida; bandits and gangs of thieves emerged from Florida to prey upon the early settlers. Pirates and smugglers worked the inland waters, and Indian attacks occurred well into the 1830s. Most permanent residents established themselves in relative security on the barrier islands."[2]

Despite the challenges, a thriving coastal trade developed. The forests of the mainland and coastal islands became a rich source of timber. Sea island cotton and rice plantations thrived in the soil and climate of the low county. As the village of St. Marys began to flourish a few miles inland on the river of the same name, it became clear that a lighthouse was needed to aid navigation through the sandy shoals of the river's estuary. In April 1802, Congress appropriated four thousand dollars for the "erection of [a] light-house on Cumberland South Point," the southernmost tip of the Cumberland Island complex. Though the appropriation was renewed periodically, keeping funds available, eighteen years passed before the amount was increased to ten thousand dollars and work began in 1820. Winslow Lewis was given the contract to construct the lighthouse, a 105-foot octagonal brick tower with a base approximately twenty-five feet in diameter tapering to approximately ten feet at its apex. At its top, a thirteen-foot-high, glass-enclosed lantern room was crowned by an eight-foot dome, bringing the total height to 126 feet. Light was provided by an array of Lewis's patent lamps. (See Figure 12B.)

At the north end of the Cumberland Island complex, maritime traffic on St. Andrew Sound had been slower to develop. Little Cumberland Island had been granted to James Habersham in 1767 and was later owned by Revolutionary War general Nathanael Greene. By the 1830s, the island, then an uninhabited wilderness, was owned by John Floyd, a prominent and wealthy landowner. With the steady growth of population and commerce around the sound and its tributaries, the Lighthouse Establishment recognized that a light on Little Cumberland Island would be a boon to navigation. In March 1837, Congress voted to appropriate eight thousand dollars "for a light-house on the north end of Little Cumberland Island." In August of that year Floyd sold six acres to the federal government as the site for its construction. A contract to build the light and associated structures was granted to Joseph Hastings of Boston in December 1837.

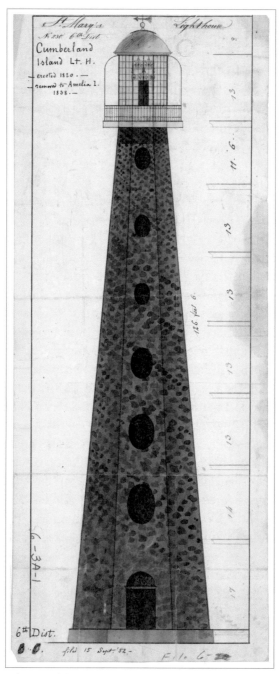

Fig. 12B Architectural drawing of Winslow Lewis's 1820 lighthouse built on the southern end of Cumberland Island. It was dismantled and its materials used to build a new lighthouse on Amelia Island, Florida.

The contract called for the new lighthouse to be a fifty-foot high conical brick tower. Its base was to be twenty-two feet in diameter and its apex eleven feet, on top of which sat an octagonal iron lantern room whose walls (except for a door to access the surrounding deck) were made of glass windows. Atop this was an iron and copper roof with a ventilator and wind vane. When finished, the lighthouse tower was to be painted white with black trim. A keeper's house, also in brick and measuring twenty by thirty-four feet in size, was included in the contract as was a privy and water well.[II] The lighting apparatus was the standard of the day, fifteen Winslow Lewis's patent lamps with sixteen-inch reflectors, initially using whale oil as fuel. The Little Cumberland light was placed in operation on June 26, 1838, emitting a steady white light. David Thompson was appointed the first keeper, earning a salary of four hundred dollars per year. He would serve in that position until 1849, continuing his other occupation as a farmer during the same time. In the early days of American lighthouses there were few, if any, restrictions on what other activities a keeper might pursue, so long as he (or she) fulfilled their primary duties at the lighthouse.[107]

While the new lighthouse was being constructed at the north end of the Cumberland Island complex, the 1820 lighthouse at the south end was being dismantled. Florida had remained a Spanish possession until 1821 when ownership was ceded to the United States.[III] In the nearly two decades since the transition in control, commerce in and out of Cumberland Sound continued to increase. The location of the lighthouse on the southern end of Cumberland Island was considered less than optimal, leading to a plan to move it across the river to a higher and more visible site on Amelia Island. While using bricks and other construction material from Cumber-

---

[II] A second keeper's dwelling was added in 1880–1881. This became the residence of the head keeper while the older 1838 dwelling became the assistant keeper's quarters.

[III] Florida became a United States territory based on the Adams-Onís Treaty, signed on February 22, 1819, with the formal transfer of ownership scheduled to take place two years later on February 22, 1821. The major provision on the part of the United States was a guarantee of the border between US territory and the colonial Viceroyalty of New Spain. The latter declared independence from Spain in August 1821, becoming the republic of Mexico. Florida was admitted to Union as the twenty-seventh state in March 1845.

land, the reconstructed lighthouse, placed into service in 1838, bears little resemblance to the original Winslow Lewis design.[IV] (See Figure 12C.)

The creation of the new Lighthouse Board in 1852 was designed to bring both managerial and technological improvement to the American system of lighthouses and other navigational aids. One of the major advances was the program to refit lighthouses with new and far more powerful Fresnel lenses, usually replacing the then commonly used Winslow Lewis lighting

Fig. 12C The lighthouse on Amelia Island, Florida was constructed in 1838 using building materials from Winslow Lewis's 1820 lighthouse on Cumberland Island.

[IV] The Amelia Island light was fully automated in 1970 and is now owned by the city of Fernandina Beach, Florida. It is occasionally open for public tours.

systems. In 1857, Little Cumberland's octagonal lantern room was removed and replaced with a brick watch room. Above this a new decagonal lantern room was installed with a French-made, third-order Fresnel lens, effectively raising the focal plane of the light to seventy-one feet above mean sea level. The following year, 1858, would be marked by an event that would leave an indelible stain on the timeline of American history, the arrival of the schooner *Wanderer* and her cargo of African slaves. She would be the penultimate slave ship to arrive in America, part of an episode that would involve the Little Cumberland lighthouse in a tangential way.[v]

THE SLAVE SHIP *WANDERER*

The *Wanderer* was built in New York in 1857 as a luxurious pleasure yacht for Col. John Johnson, "a wealthy sportsman and member of the New York Yacht Club, who had spared no expense in making her one of the finest pleasure craft in the world."[4] A sleek, two-masted schooner, the vessel was said to be able to reach speeds of up to twenty knots, equaling the speed of some of the fastest ships of the day. In early 1858, the *Wanderer* was sold to William C. Corrie of Charleston, South Carolina, who, in collaboration with Charles Augustus Lafayette Lamar of Savannah, planned to use the vessel to import slaves from Africa. A wealthy businessman and member of a prominent family, Lamar was a vocal advocate of the resumption of the slave trade. In this, the Golden Age of the South's cotton-based economy, slave labor was considered an indispensable necessity, and therefore a rich source of revenue to those like Lamar who were willing to defy the law.

    The importation of African slaves into the United States was no easy task. It had been illegal since January 1, 1808, based on an act of Congress passed in March 1807. The federal government was quick to prosecute any accused of the crime, and both American and British fleets were stationed along the coast of west Africa to

---

[v] A detailed account of the *Wanderer* episode is contained in Tom H. Wells's *The Slave Ship Wanderer*, originally published by the University of Georgia Press in 1967, and reprinted in a softcover edition in 2009. Many of the details in this account are based on that reference.

Fig. 12D "Wanderer" with her cargo of African slaves arrived at St. Andrew Sound on November 28, 1858.

intercept would-be slavers. The profits to be made, however, were enormous, justifying the risks in the minds of those willing to take the chance.

The *Wanderer* was refitted to be able to carry its human cargo and departed Charleston under the command of Corrie on July 3, 1858, sailing first to the island of Trinidad, then eastward toward Africa. Arriving at the mouth of the Congo River on September 15th, Corrie and his crew spent the next several weeks in the area negotiating the purchase of slaves while avoiding impoundment by both American and British warships. In mid-October, the *Wanderer*, with a human cargo of nearly five hundred slaves, departed for North America.[VI]

The transatlantic voyage took about six weeks. The plan was to unload the slaves on Jekyll Island, then owned by members of the Dubignon family and managed as a plantation for the production of cotton, corn, and related crops. The island, private and easily

---

VI Estimates vary, but based on later testimony and the ship's log, the *Wanderer* departed Africa with about 490 slaves.

accessible from St. Andrew Sound, offered a discreet location to offload the illicit cargo. Thereafter, the slaves were to be quietly sold in nearby Georgia and South Carolina markets.

The *Wanderer* arrived off St. Andrew Sound on November 28, 1858. As the hidden shoals and shifting currents made entry into the sound dangerous without a local pilot, near sunset Corrie and another man rowed ashore in a small boat to the Little Cumberland lighthouse in search of someone to help guide the vessel to the Jekyll landing. There they found Horatio Harris, who identified himself as the substitute lighthouse keeper. The regular keeper, James A. Clubb, was said to be visiting across the sound on Jekyll Island. Importantly, Clubb was described as a "retired pilot" and able to bring the *Wanderer* in safely if he would agree to do so. The three rowed across the sound and eventually found Clubb, who was visiting with John Dubignon, a member of the family that owned the island.[5]

The rumors that the *Wanderer* was a slave ship were rampant, and Clubb was at first reluctant to be of assistance. Later, deciding that acting as pilot "was a money making piece of business and thought he had a right to come in for a share," Clubb said he would do the piloting for five hundred dollars, more than a year's lighthouse keeper's income for a job which under normal circumstances would have cost fifteen to nineteen dollars.[6] In the early morning darkness of November 29th, Clubb brought the yacht into St. Andrews Sound and safely dropped anchor about two hundred yard off the Dubignon landing facing the marsh on the south end of Jekyll Island. The Africans were unloaded and turned over to Dubignon. The exact number that had died during the transatlantic voyage is unknown; the number of 409 survivors who would become American slaves is widely quoted. Over the following weeks, most of the new slaves would be moved to the Savannah River area to be sold, with others sold in the local markets around Jekyll Island. With the schooner's human cargo unloaded, Clubb took the *Wanderer* two miles up the Little Satilla River, where he anchored it so as not to been seen by other ships on the sound.

Even though the *Wanderer's* owners and others directly involved had taken precautions to keep the true purpose of the voyage to Africa a secret, word of the sudden influx of new slaves and their presumed origin spread rapidly. The following months saw a series of legal actions including indictments and trials. Despite Charles Lamar's offer of five-thousand-dollar bribes to John Clubb and Horatio Harris not to testify, both agreed to serve as witnesses for the state in the prosecution of Lamar, the Dubignon brothers, and several others for the crimes related to the importation of slaves. In the end there were no substantial convictions or fines. This was in large part due to the increasing schism between North and South over the issue of slavery. The eventual fate of what Congress had decreed a crime half a century earlier would soon be settled by armed conflict.

After being impounded as a slave ship, the *Wanderer* was sold at auction to Charles Lamar in March 1859. She would eventually be seized and serve the United States Navy during the Civil War. At the end of hostilities, she was sold into mercantile service in June 1865, and lost at sea near Cuba in January 1871.[VII]

THE CIVIL WAR YEARS AND THE DEATH OF A UNION SAILOR

On January 19, 1861, the Georgia Secession Convention, meeting in the capital of Milledgeville, voted to secede from the Union. In some respects, the transition was initially seamless, with control of former federal institutions transferred first to the state and subsequently to Confederate States institutions as appropriate. The Confederate Lighthouse Bureau was created in March 1861 but quickly became irrelevant with the attack on Fort Sumter and Lincoln's institution of the Southern naval blockade the following month.

Along Georgia's coast, there were initial efforts to guard sea approaches to the major cities and population centers, including a Confederate artillery battery placed on Little Cumberland Island facing St. Andrew Sound. It became increasingly clear, however, that many of the coastal areas were indefensible. By the spring of

---

[VII] Though far less well documented, the last slave ship to arrive in America is said to be the *Clotilda*, which arrived in Mobile Bay, Alabama, in 1859 or 1860. A monument commemorating the *Wanderer* episode can be viewed near the St. Andrew Picnic Area near the southern end of Jekyll Island.

1862, most defensive positions had been withdrawn. On March 2, 1862, Union warships sailed unimpeded into St. Andrew Sound, noting that the American flag was flying atop the Little Cumberland lighthouse. Federal troops had earlier discovered that the lighthouse had been abandoned, its lens apparatus removed and allegedly sent to Brunswick for storage.[VIII] The light would remain dark for the remainder of the conflict.

The war years passed without major events on Little Cumberland Island. The blockading fleet spent much of its time attending to issues of refugees, both freed slaves as well as members of the local populace. Among the ships that rotated blockade duty on St. Andrew Sound was the bark *Braziliera*, a wooden sailing ship stationed there from June 1863 to August 1864.

Fig. 12E By March 1862, the Little Cumberland lighthouse had been abandoned, its lens removed by the Confederates. Federal troops erected the United States flag as Union naval forces sailed into St. Andrew Sound.

On the morning of November 21, 1863, three sailors, including Quartermaster Charles R. Farnham, left the *Braziliera* in a small boat to take up picket duty on Jekyll Island. Several hours later it was reported that the boat carrying the three had swamped as it approached the beach. Farnham disappeared into the surf and was presumed to have drowned. He was a brown-haired, fair-complected, twenty-one-year-old from Weymouth, Massachusetts, who formerly worked as a harness maker before enlisting in the navy in January 1863. An immediate search failed to recover his body, but a funeral service was held aboard ship the following morning. A further search two days later recovered his remains. He was buried on the afternoon of November 24th just north of the

---

VIII A search was made for the missing Fresnel lens, but it was never recovered.

Little Cumberland lighthouse. With time, Farnham's grave would disappear and be lost under the constantly shifting sand dunes, only to be discovered more than a century later.[IX]

The end of the war brought some degree of peace to coastal Georgia, although the social and political changes wrought by the conflict would linger for decades. Damages to the Little Cumberland lighthouse, keepers' cottages, and related structures, many of which may have resulted from more than six years of neglect, were repaired by mid-1867. A new, third-order Fresnel lens was installed and lighted for the first time on September 1, 1867.

THE LAST HALF CENTURY

The Reconstruction Era (1865–1877) brought some degree of stability to Georgia, so recently ravaged by the Civil War and the vast sociopolitical changes brought by the abolition of slavery and the end of the plantation system. Commerce, however, soon started a steady recovery propelled by the reborn cotton economy based on tenant farming and the explosive growth of the naval-stores industry in Georgia's wiregrass country inland from the coast. Records of the Lighthouse Bureau in 1870 note concern that the sea was encroaching on the Little Cumberland light, prompting in 1876 the construction of a retaining wall near the tower's base.

By the 1880s, the sunny coast of Georgia was attracting visitors from the North. Jekyll Island, for example, just across St. Andrew Sound from Little Cumberland, was sold in 1886 by the Dubignon family to the Jekyll Island Club, whose members included Henry Hyde, Marshall Field, John Pierpont Morgan, Joseph Pulitzer, and William K. Vanderbilt. On a more modest scale, accommodations on greater Cumberland Island were available to the public. Writing to her hometown newspaper in May 1882, Angele Davis, a visitor from Cincinnati, described her party's visit to Little Cumberland Lighthouse:

[IX] At the end of the conflict, the Lighthouse Bureau assumed responsibility for the maintenance of Farnham's grave. National Archives records include an 1894 invoice from the bureau to the navy requesting forty dollars for the erection of a marble headstone and maintenance of the fence surrounding the burial site.

We landed on the white glaring beach of Little Cumberland, and, leaving the boat, started for the light-house. We had gone only a short distance when we met the assistant light-keeper. The man is a character. He was born in Pennsylvania, and coming South before the war, entered the Confederate service. He served four years, and after many vicissitudes, obtained the position of assistant keeper at Cumberland.

He married a Spanish-American woman from St. Augustine. This wife of his shoots in a way calculated to make an average sportsman green with envy. She killed a dozen rice birds at one shot the day before we were here. There are droves of hogs running almost wild on the island. And this modern Diana[X] takes a gun, follows them up, kills one and brings him home on her shoulder. She is a slender woman too, but her muscles must be made of steel.

The light-house and the houses are in a thicket of low jack oaks that hide the sea from the dwellings. Back of them is a beautiful grove. This thicket of oaks is alive with mockingbirds that sing all the hours of day and night. They are always undisturbed, and almost as tame as chickens.... The thing that struck me most forcibly was the cleanliness of everything. That lighthouse is certainly the most spotless place that has yet been discovered. Everything is painted, whitewashed, scrubbed and scoured to the uttermost extent. The light is a stationary light. The one just above, on St. Simons Island, is a flash light.... We went out on the little iron balcony just outside the light. It is a slippery perilous place, and I could not help thinking of the consequences if my feet should slip.[7, XI]

The Little Cumberland lighthouse survived the Charleston earthquake of August 31, 1886, with only minor damage, and the massive hurricanes of 1893, 1895, and 1898 that left a path of death and destruction along the Georgia coast. Each time repairs were made to the lighthouse and associated structures. With conversion of the lighting fuel to kerosene ("mineral oil"), a separate brick oil

---

[X] Diana was a Roman goddess, the patroness of hunters and the hunt.

[XI] The unnamed assistant lighthouse keeper was Isaac L. Peckham, who would be appointed head keeper at the St. Simons lighthouse in January 1883 at a salary of $600 per year. His wife was appointed assistant keeper the same year at an annual salary of $400. Both would serve until their retirement in 1893.

house was built in 1890, identical to ones built the same year at Tybee, Sapelo, and St. Simons lighthouses. An Endicott Era gun emplacement, together with another at the southern end of Jekyll Island, was briefly located near the lighthouse for a few months in 1898. The artillery batteries were placed to protect St. Andrew Sound from Spanish attack following the sinking of the *USS Maine* and the four-month-long Spanish-American War.[XII]

With advances in navigational science and changes in patterns of commerce and trade, the relative value of the Little Cumberland lighthouse decreased, concomitant with increasing costs of personnel and maintenance. In 1914, the Lighthouse Bureau declared the light obsolete. It was extinguished on March 15, 1915.

THE SECOND LIFE OF LITTLE CUMBERLAND LIGHTHOUSE

For most of the next half century, the Little Cumberland lighthouse remained abandoned. In 1923, the six-acre tract on which it stood was sold by the government to a Brunswick real estate firm. Over the following decades, the island, including the lighthouse tract, changed owners several times. During the years of Prohibition, the island became a haven for bootleggers and rum-runners. Later owners hoped to develop it for tourism purposes, but none of their plans came to fruition. In 1961, the island was acquired by the Little Cumberland Island Homes Association, Inc. (LCIHA), whose stated mission was "to preserve the island and its natural integrity to the extent possible while providing a limited number of second home sites for Association members."[8] The initial plan called for up to ten percent of the land to be developed as cottages for members, with the remaining to be left as wilderness. To this day, the association has maintained those guidelines. Though privately owned, under a preservation agreement with the National Park Service, Little Cumberland Island is part the Cumberland Island National Seashore.

XII See Chapter 11 for more information on Endicott Era artillery emplacements.

In 1967, the association turned its attention to the long-neglected lighthouse with a plan to restore it to the extent possible. A twenty-foot-high sand dune blocked entrance to the tower's door and completely covered the adjacent 1890 oil house. "Sand from the dune had pushed through the lowest of the three tower windows and packed the ground floor of the tower."[9] A survey reached the

Fig. 12F A photo taken of the Little Cumberland lighthouse in the mid-1960s prior to initial restoration efforts. The original 1838 construction makes up the lower part of the tower. Above that can be seen the 1857 watch room and the remains of the badly deteriorated lightroom of the same era. Below, a sand dune twenty feet in height is blocking the tower's entrance.

Fig. 12G The remains of the Little Cumberland lighthouse keepers' cottages are seen in this photo from the mid-1960s. The original brick 1838 cottage is in the background. The structure to the left is the porch of the 1880–1881 second keeper's cottage.

conclusion that the oil house and the remains of the keepers' quarters were beyond repair. To raise funds to begin the renovation of the tower, the bricks from those structures were sold to association members for use in the construction of their island homes. While the basic masonry of the tower remained intact, interior water damage and deterioration of the lightroom and superstructure was significant. Over the following months, the first steps were taken to save the lighthouse. The sand was removed from the tower and the steps rebuilt. The lantern room was rebuilt and windows installed to help prevent further water damage to the interior.

In the early 1990s, Camilla Merts, a property owner and member of the Little Cumberland Island Homes Association, made it her mission to save the lighthouse. Assisted by leadership teams and members focusing on the lighthouse plus a succession of island operation managers, the lighthouse has been maintained. Working first with a historic-preservation architect, the tower was given a fresh coat of whitewash formulated to duplicate the original coating from

Fig. 12H The 1863 grave of Union sailor Charles Farnham (red arrow) can be seen emerging from an eroded sand dune near the Little Cumberland lighthouse.

1838. The lantern room windows and roof, which had continued to leak, were replaced. In the midst of the restoration project, an unexpected discovery led to an additional task for the team.

To the north of the lighthouse a series of sand dunes lay between the tower and the beach. In October 1994, active erosion of the beachfront dune revealed portions of an ornamental iron fence protruding from the sand. On further investigation this was found to be the grave of Union sailor Charles Farnham, buried there in 1863. Over the years the grave, with its fence and headstone, had been covered by more than six feet of windblown and surf-driven sand. With the current dune erosion, it was now in danger of collapsing onto the beach. With the assistance of an archaeologist, Farnham's remains as well as the fence and headstone were relocated to a more secure site approximately one hundred yards southwest of the lighthouse.[10]

By 2009 the preservation work done more than a decade earlier had begun to deteriorate, necessitating additional repairs and a long-term preservation plan. Working with a nationally recognized lighthouse expert and another architect specializing in lighthouse preservation, the group began a further round of repairs. Over the next five years, the lantern room was reaffixed to the tower, the exterior was cleaned and recoated, and the brick tower itself restrengthened. The project was completed in August 2015 and celebrated with an open house in January 2016.

With firm commitment and a long-term preservation plan now in place, the Little Cumberland lighthouse, now more than 180 years old, can look forward to its third century of existence.

A NOTE TO WOULD-BE VISITORS

Little Cumberland Island remains for the for the most part wilderness, its dune-shaped landscape covered with forests of live oak, pine, and palmetto, surrounded by the sea and tidal creeks. Though crisscrossed by well-maintained all-terrain-vehicle paths, obvious evidence of human impact is minimal. Bird life is abundant, with traditional seabirds as well as ospreys and bald eagles. Sea turtles nest on its white-sand beaches on which can be seen the occasional feral horse from among the several that migrated over from nearby Cumberland Island. It is a living monument to ecological preservation, which is one of the prime goals of its owners.

The Little Cumberland lighthouse exists today because of the commitment and dedication of members of the Little Cumberland Island Homes Association. To quote one member of the lighthouse preservation team, "The lighthouse has become our icon—it's the first thing we see on approaching the island and the last thing we see as we depart. We preserve it because we love it and spend our hard-earned money to do this. We respectfully request that all enjoying this book and this chapter on our lighthouse will also respect our privacy and not trespass. Our island is private; we maintain it only with our private funds. We do our utmost to uphold our preservation agreement with the Department of the Interior. Please enjoy the lighthouse as described in this book. Please enjoy it from a boat, but please do not trespass."

APPENDIX

A BRIEF GLOSSARY OF LIGHTHOUSE-RELATED TERMS

Arago Lamp (also Fresnel-Arago Lamp)
A variant on the oil lamp of Ami Argand which employed four separate concentric wicks around a central airflow shaft to produce a flame that was said to be twenty times brighter than any other light source available at the time. It was invented around 1820 by Augustin Fresnel working with French physicist François Arago as the two sought a brighter type of lamp for use in lighthouses.

Argand Lamp
A new, and at the time revolutionary, type of oil lamp invented by Swiss researcher Ami Argand in the 1780s. Through the use of a circular wick and a chimney to increase the flow of air, the Argand lamp produced a flame that was far brighter than previous oil lamps while being far more efficient.

Beacon
Most commonly, a light or other bright display capable of serving as a navigational guide. Although the word "beacon" in current usage implies the presence of a light, in older terminology it often referred to a prominent landmark or signal tower. A beacon can also refer to an invisible guide, as in a radio beacon.

Bullseye Lens
A convex lens that is used to focus light via refraction.

Catadioptric
An adjective that refers to light that has been both *reflected* and *refracted*.

Catoptric
An adjective that refers to *reflected* light. The word is derived from *katoptron*, the Greek word for mirror.

Opposite, Tybee Island lighthouse.

Daymark
A *daymark* is a prominent land-based object or structure serving as an aid to navigation that can be identified during daylight hours, i.e., a landmark. *Seamark* is the equivalent term for a conspicuous object distinguishable *at or from* the sea serving as a navigational guide or to warn mariners of danger. Seamarks can either be on land, e.g., lighthouses, or at sea, e.g., buoys or lightships. Rephrased, a landmark can be a seamark, but a seamark is not necessarily a landmark.

Diffraction
The process by which a beam of light or other system of waves is spread out as a result of passing through a narrow aperture or across an edge, typically accompanied by interference between the wave forms produced.

Dioptric
An adjective that refers to light whose path has been altered by passage through a lens or similar. An equivalent term is *refracted*.

Fresnel Lens
A type of lens invented by French civil engineer and physicist Augustin Fresnel in the early 1820s. At the time, Fresnel was seeking a better and more efficient way to project beacons from lighthouses. By taking advantage of the reflection and refraction produced by a convex lens and prisms of glass, light is intensified and focused in a single direction. The practical invention of this type of lens represented a revolution in optics. Fresnel lenses today are in common use in multiple applications ranging from vehicle lights to camera lenses to airport landing lights. Though formerly manufactured using precisely cut and polished glass, many are now formed in plastic.

Geographic Range
Refers to the approximate distance at which an object may be seen by an observer at sea level, assuming maximal visibility of meteorological conditions.

Harbor Light
A light that is available to guide ships safely into a harbor.

Knot (as a measure of distance)
One nautical mile per hour. (See "Nautical Mile.")

Lantern Room
Usually the uppermost enclosed space atop a lighthouse tower, commonly enclosed by glass windows and housing the lighting apparatus.

Lewis Lamp
A type of light fixture patented by Winslow Lewis in 1810 and commonly used in American lighthouses from 1812 until 1852. Although Lewis claimed this to be his invention, it borrowed heavily from others, using a modification of the Argand lamp, a reflector, and a bullseye lens to focus the light. Lighthouse beams, generated by arrays of Lewis lamps, were considered vastly inferior to the contemporary Fresnel lens system.

Lightship
A ship that served as a water-based lighthouse, usually employed in areas where construction of a lighthouse was not possible. Lightships were usually firmly anchored in a fixed position, their light sources elevated high above the sea on masts.

Littoral
As a noun, referring to a coastal region, e.g., the Venezuelan littoral. As an adjective, referring to the coast of a body of water, e.g., a littoral zone, or littoral warships.

Luminous Range
Refers to the distance at which a light may be expected to be seen when adjusted for prevailing meteorological conditions.

Meteorological Visibility
A scale based on the International Visibility Code that rates the transparency of the air for the transmission of light from 0 to 9, with 0 being less than 50 yards to 9 being greater than 27 nautical miles.

Nautical Mile
A unit of distance used at sea. A nautical mile is defined as the average distance on the Earth's surface along one minute of latitude (1/60 of a degree) along any line of longitude. The international nautical mile equals 1.1508 statute miles or 1,852 meters.

Nominal Range
The maximal distance that a lighthouse beam can be seen in clear weather.

Occulting Light
A light in which the total duration of light in each period is longer than the total duration of darkness and in which the intervals of darkness (occultations) are all of equal duration. Occultations are created by partially blocking, or occulting, the light to make it appear to flash.

Prism
An optical prism is a transparent (or translucent) element with polished surfaces that serve to refract light. At least one surface must be at an angle to another. Prisms are commonly made of glass or plastic, though other substances may be used.

Range Lights (or Range Beacons)
At least two beacons, visible by day and illuminated by night, that are used as navigational aids in maintaining a ship's bearing. The nearer beacon, referred to as the front range light, is commonly placed at a lower elevation that the other (the rear range light). By visually lining up the two and maintaining their alignment, a vessel's pilot can be certain he/she is on a proper course.

Reflection
As applied to visible light, the throwing back of light waves by an object or surface.

Refraction
As applied to visible light, the bending of light waves as they pass from one translucent substance to another.

Roadstead (or Roads)
Generally speaking, a roadstead is a site where a ship can anchor reasonably safely without fear of tides or currents causing the anchor to drag or be displaced. River estuaries are common sites of roadsteads. Such sites were particularly important in the days of sailing ships, when a vessel might have to wait for favorable winds to begin its journey. In Georgia, for example, the area just off Tybee Island where the Savannah River empties in the sea is known as Tybee Roads. "In maritime law, a known general station for ships, notoriously used as such, and distinguished by the name; and not any spot where an anchor will find bottom and fix itself." (Black's Law Dictionary)

Seamark
A conspicuous object visible from the sea, usually serving to guide mariners or warn them of danger. Seamarks can either be on land, e.g., lighthouses, or at sea, e.g., buoys or lightships.

GLOSSARY

Spermaceti Oil (or "Sperm Oil")
The term "spermaceti oil" is a bit of a misnomer. Spermaceti is a waxy liquid substance created by the spermaceti organ and found in the head cavities of sperm whales. A large whale may contain up to five hundred gallons. During the peak whaling years of the nineteenth century, whalers would extract the spermaceti and initially process it on board their vessels before storing it in casks for further processing on land. "Winter-strained sperm oil" had the advantage of remaining liquid during freezing winter temperatures, an important asset for lighthouse illumination. Spermaceti wax, another product of spermaceti, was at one time widely used for candle making, in cosmetics, and as a lubricant.

Watch Room
A room near the top of a lighthouse tower, generally just below the lantern room, where a lighthouse keeper could stay during hours of darkness to attend the light.

SUGGESTED FURTHER READING

Seemingly innumerable books have been written on lighthouses, both in the genres of fiction and nonfiction. The brief list that follows contains references to a limited number of books and digital resources that address lighthouse history, technology, and related subjects. Printed material can be easily found by an online search.

GENERAL BOOKS ON LIGHTHOUSES AND RELATED TOPICS:

Brilliant Beacons by Eric Jay Dolan (2016): A relatively recent general overview of lighthouses, including their history, construction, and technology.

Lighthouses and Lightships of the United States by George R. Putnam (1917): An older but in some areas more detailed overview of American lighthouses written by the man who would serve as American commissioner of lighthouses from 1910 to 1935. The book is now in the public domain; digital copies can be downloaded from the internet.

Sentinel of the Coasts—The Log of a Lighthouse Engineer by George R. Putnam (1937): Putnam's autobiography is both an interesting story and a detailed account of his challenges and successes during his tenure as lighthouse commissioner.

America's Lighthouses—An Illustrated History by Francis R. Holland Jr. (1972): An excellent and often detailed history of lighthouses of the United States, with many unique photographs, diagrams, and illustrations.

The World's Lighthouses Before 1820 by D. Alan Stevenson (1959): A history of lighthouses focusing mainly on Europe, written by one of the members of the famous Stevenson family of lighthouse architects and builders. This aspect is important in that American lighthouses were based on European designs and lighted by European technology. A reprint of this work was released in 2002.

A Short Bright Flash by Theresa Levitt (2013): In part an excellent history of Augustin Fresnel and his development of lighthouse optics, and in part an overview of the development of the American lighthouse network.

Longitude by Dava Sobel (1995): A very readable account of the discovery of an accurate way to calculate longitude, perhaps the greatest problem facing mariners through the late eighteenth century.

ONLINE RESOURCES:

The United States Lighthouse Society (www.uslhs.org): One of the very best digital resources on American lighthouses, with explanatory articles, diagrams, photos, and numerous links to sources of information. Per its website, "The United States Lighthouse Society is a nonprofit historical and educational organization dedicated to saving and sharing the rich maritime legacy of American lighthouses and supporting lighthouse preservation throughout the nation."

The National Archives, Record Group 26 (https://www.archives.gov/findingaid/stat/discovery/26): The United States National Archives is the repository of many records relating to lighthouses dating back to the late eighteenth century, in the early days of the republic. Record Group 26 contains records related to the Coast Guard, which has managed American lighthouses since 1939. The database, which contains maps, photographs, architectural drawings, log books, correspondence, and more, is easily searchable by topic or keyword.

IMAGE CREDITS

Chapter 1: **The Mystique of Lighthouses**
Fig.1A From *Lighthouses*, by David Stevenson (Edinburgh, SCT: Adam & Charles Black, 1865).

Chapter 2: **Lighthouses through the Ages**
Fig. 2A Emad Victor Shenouda, Wikipedia; **Fig. 2B** Chris McKenna, Wikipedia; **Fig. 2C** Public Domain

Chapter 3: **Lighthouse Construction**
Fig. 3A *Encyclopedia Britannica*, Eleventh Edition, 1911; **Fig. 3B** Public Domain; **Fig. 3C** *Encyclopedia Britannica*, Eleventh Edition, 1911; **Fig. 3D** *Encyclopedia Britannica*, Eleventh Edition, 1911

Chapter 4: **Throwing the Light**
Fig. 4A From Hardy, W. J., *Lighthouses: Their History and Romance* (Oxford: The Religious Tract Society, 1895); **Fig. 4B** Public Domain; **Fig. 4C** Reproduced with permission of the U.S. Lighthouse Society, illustration by Jim Burt; **Fig. 4D** Reproduced with permission of the U.S. Lighthouse Society, illustration by Jim Burt; **Fig. 4E** Reproduced with permission of the U.S. Lighthouse Society, illustration by Jim Burt; **Fig. 4F** Public Domain; **Fig. 4G** Courtesy Bibliotheque Centrale de l'École Polytechnique; **Fig. 4H** From Ganot, Adolphe, *Natural Philosophy*, English translation, 6th edition (New York: D. Appleton and Company, 1887); **Fig. 4I** From *Vanity Fair*, December 31, 1869; **Fig. 4J** From *Encyclopedia Britannica*, Eleventh Edition, 1911

Chapter 5: **American Lighthouses**
Fig. 5A Courtesy Library of Congress; **Fig. 5B** Courtesy Library of Congress

IMAGE CREDITS   191

Chapter 6: **Lighthouses during the American Civil War**

Fig. 6A Public Domain; **Fig. 6B** Courtesy National Archives; **Fig. 6C** Courtesy Library of Congress; **Fig. 6D** Courtesy National Archives; **Fig. 6E** Courtesy Library of Congress

Chapter 7: **Keepers of the Light**

Fig. 7A From *The Keeper's Log*, Fall 2001, reproduced with permission of the U.S. Lighthouse Society; **Fig. 7B** From *The Literay Digest* and Funk & Wagnalls Company, 1923; **Fig. 7C** Courtesy William Rawlings; **Fig. 7D** Courtesy Dundee Art Galleries and Museums Collection (Dundee City Council); **Fig. 7E** From *Harper's Weekly*, July 31, 1869; **Fig. 7F** From *Frank Leslie's Illustrated Newspaper*, November 5, 1881

Chapter 8: **Tybee Lighthouse**

Fig. 8A Courtesy William Rawlings; **Fig. 8B** Courtesy National Archives; **Fig. 8C** From Frank Leslie's *The Soldier in Our Civil War*, vol. 1 (New York & Atlanta: Stanley Bradley Publishing Company, 1893); **Fig. 8D** Courtesy National Archives; **Fig. 8E** Courtesy National Archives; **Fig. 8D** Courtesy National Archives; **Fig. 8G** Courtesy William Rawlings; **Fig. 8H** Courtesy Tybee Island Historical Society

Chapter 9: **Cockspur Lighthouse**

Fig. 9A Courtesy National Archives; **Fig. 9B** Courtesy National Archives; **Fig. 9C** Courtesy Library of Congress, Historic American Buildes Survey, ca. 1994; **Fig. 9D** Courtesy Library of Congress, Historic American Buildes Survey, ca. 1994; **Fig. 9E** Courtesy Library of Congress; **Fig. 9F** Courtesy Library of Congress; **Fig. 9G** Courtesy William Rawlings; **Fig. 9H** Courtesy William Rawlings; **Fig. 9I** Courtesy William Rawlings

Chapter 10: **Sapelo Lighthouse**

Fig. 10A Courtesy William Rawlings; **Fig. 10B** Courtesy National Archives; **Fig. 10C** Courtesy National Archives; **Fig. 10D** Courtesy William Rawlings; **Fig 10E** National Archives: From National Register of Historic Places Application; **Fig 10F** Courtesy William Rawlings; **Fig. 10G** Courtesy,Georgia Archives, Vanishing Georgia Collection, Image SAP012; **Fig. 10H** National Archives: From National Register of Historic Places Application; **Fig. 10I** Courtesy William Rawlings

Chapter 11: **St. Simons Lighthouse**
Fig. 11A Courtesy Coastal Georgia Historical Society; **Fig. 11B** Courtesy National Archives; **Fig. 11C** Courtesy National Archives; **Fig. 11D** Courtesy National Archives; **Fig. 11E** Courtesy National Archives; **Fig. 11F** Courtesy National Archives; **Fig. 11G** Courtesy Coastal Georgia Historical Society; **Fig. 11H** Courtesy Willilam Rawlings; **Fig. 11I** Courtesy Coastal Georgia Historical Society

Chapter 12: **Little Cumberland Lighthouse**
Fig. 12A Courtesy William Rawlings; **Fig. 12B** Courtsey National Archives; **Fig. 12C** Courtesy National Archives; **Fig. 12D** From *Harper's Weekly*, January 8, 1859; **Fig. 12E** From *Frank Leslie's Illustrated Newspaper*, March 29, 1862; **Fig. 12F** Courtesy members of the LCIHA; **Fig. 12G** Courtesy members of the LCIHA; **Fig. 12H** Courtesy members of the LCIHA

ENDNOTES

Chapter One

1. From the *Northampton* (MA) *Courier*, quoted in the *Daily Herald* (New Haven, CT) 19 December 1839.
2. *Cabinet* (Schenectady, NY) 24 December 1839.
3. *Boston Recorder*, 31 January 1840.
4. *Daily Pennsylvanian* (Philadelphia), 13 January 1840.
5. *Observer* (New York), 4 January 1840.
6. Ibid.
7. Boston *Gazette*, quoted in the Gloucester *Telegraph*, 4 January 1840. Emphasis and spelling are per the original.
8. *Hugh Tennent v. The Earl of Glasgow* (1864) 2 Paterson 1229, https://www.casemine.com/judgement/uk/5a8ff8c760d03e7f57ecd345.
9. *https://Thebookcoverdesigner.com* (accessed 19 June 2019).
10. *https://safepassageurns.com/pages/lighthouse-symbolism* (accessed 19 June 2019).
11. *https://thelighthousefortheblindinc.org/* (accessed 19 June 2019).

Chapter Two

1. Dava Sobel, *Longitude* (New York: Walker and Company, 1995) 15; Alan D. Stevenson, *The World's Lighthouses from Ancient Times to 1820* (Mineola, NY: Dover Publications, 2002 reprint of 1959 edition) 16.
2. Stevenson, *World's Lighthouses*, xxiii.
3. Adapted from Stevenson, *World's Lighthouses*, 87.

Chapter Three

1. Adapted from Stevenson, *World's Lighthouses*, 113.

Chapter Four

1. Eric J. Dolan, *Brilliant Beacons* (New York: Liveright Publishing Corporation, 2016) 71.
2. Bella Bathhurst, *The Lighthouse Stevensons* (New York: Harper Collins Publishers, 1999) 46.
3. For a more detailed history of the story of the Statue of Liberty from the perspective of its use as a lighthouse, see "Liberty Enlightening the World," parts 1 and 2, by Carole L. Perrault, published in *The Keeper's Log* (Spring and Summer 1986 issues). A brief biography of Frédéric-Auguste Bartholdi as it pertains to the Statue of Liberty can be found on the National Park Service Statue of Liberty website, https://www.nps.gov/stli/learn/historyculture/frederic-auguste-bartholdi.htm (accessed 9 December 2019). Quotes in this section are drawn from these sources.
4. See "The Use of Modern Light Sources in Traditional Lighthouse Optics," International Association of Lighthouse Authorities (IALA) Guideline 1049, Edition 2.0, December 2007; and "Light Sources Used in Visual Aids to Navigation," International Association of Lighthouse Authorities (IALA) Guideline 1043, Edition 1.2 December 2011.
5. Ibid.

Chapter Five

1. http://tps.cr.nps.gov/nhl/detail.cfm?ResourceId=1738019993&ResourceType=Structure (accessed 5 July 2019).
2. https://en.wikisource.org/wiki/United_States_Statutes_at_Large/Volume_2/12th_Congress/1st_Session/Chapter_34 (accessed 5 July 2019).
3. Francis R. Holland Jr., *America's Lighthouses: An Illustrated History* (New York: Dover Publications, 1972) 27.
4. Ibid.
5. https://uslhs.org/history-administration-lighthouses-america (accessed 3 July 2019).
6. Quoted in *Appleton's Cyclopedia and Register of Important Events*, vol. 5 (New York: D. Appleton & Co., 1886) 437.
7. *Statutes at Large and Treaties of the United States of America 1850–51* (Boston: Little & Brown, 1851).
8. Ambrose Bierce, *The Devil's Dictionary* (New York: World Publishing Company, 1911) 193.
9. *Sentinel of the Coasts* (New York: W. W. Norton & Company, 1937) 127–28. See also Holland, *America's Lighthouses*, 40–41.

10 Modified from George R. Putnam, *Lighthouses and Lightships of the United States* (Boston: Houghton Mifflin Company, 1917) 52.
11 *Evening Star* (Washington, DC), 9 May 1939.
12 https://www.gps.gov/systems/gps/performance/2018-GPS-SPS-performance-analysis.pdf (accessed 11 July 2019).
13 *Los Angeles Times*, 19 December 2003.

Chapter Six
1 Raphael Semmes, *Memoirs of Service Afloat During the War Between the States* (Baltimore, MD: Kelly Piet & Company, 1869) 80.
2 Ibid., 92.
3 *Augusta* (GA) *Chronicle*, 12 September 1861.
4 Ibid., 13 September 1861.
5 *Frank Leslie's Illustrated Newspaper* (NY), 2 November 1861.
6 Rodney E. Dillon Jr., "'A Gang of Pirates:' Confederate Lighthouse Raids in Southeast Florida, 1861," *The Florida Historical Quarterly* 67 (1989): 441–57.
7 Ibid.
8 Ibid.
9 *Frank Leslie's Illustrated Newspaper* (NY), 21 September 1861.
10 Ibid., 14 September 1861.
11 Ibid., 3 January 1863.
12 Ibid., 23 April 1864.
13 From the *Raleigh Confederate* as quoted in the *Macon* (GA) *Telegraph*, 8 November 1864.
14 *New York Herald*, 6 October 1865.

Chapter Seven
1 "The Lighthouse Keeper," *Cassell's Magazine*, 9 March 1867.
2 Holland, *America's Lighthouses*, 39.
3 Ibid., ix.
4 Dolin, 208.
5 Quoted in "The Keeper's New Clothes," *The Keeper's Log* (Fall 2001).
6 Dolin, 225–26.
7 See "The Keeper's Library," *The Keeper's Log* (Summer 1995).
8 It should be appreciated that the vast majority of African Americans (approximately 90 percent) lived in the former Confederate states through the end of the nineteenth century. For more infor-

mation see "More Black Lighthouse Keepers," by N. E. Hurley and C. Belcher, in *The Keeper's Log* (Spring 2019).
9. *Boston Journal*, 6 December 1897.
10. Mary L. Clifford and J. Candace Clifford, *Women Who Kept the Lights*, 3rd edition (Alexandria, VA: Cypress Communications, 2013).
11. *San Francisco Bulletin*, 9 January 1879.
12. *Plain Dealer* (Cleveland, OH), 3 April 1886.
13. *Jackson* (MI) *Citizen Patriot*, 17 March 1890.
14. *San Francisco Bulletin*, 28 July 1891.
15. *Jackson* (MI) *Citizen*, 23 November 1900.
16. There are numerous sources and accounts of the Grace Darling story, including books and newspaper accounts throughout the nineteenth century. Currently, the most extensive and seemingly accurate is "The Grace Darling Website—Legendary Victorian Heroine" (*www.GraceDarling.co.uk*).
17. Ibid. (accessed 3 December 2019).
18. https://www.publishersweekly.com/978-0-06-269862-9 (accessed 4 December 2019).
19. *Harper's Weekly*, 31 July 1867.
20. *Frank Leslie's Illustrated Weekly*, 5 November 1881.
21. Reported in the *Plain Dealer* (Cleveland, OH), 29 September 1875.

Chapter Eight
1. Milton B. Smith, "The Lighthouse on Tybee Island," *The Georgia Historical Quarterly* 49 (1965): 246.
2. Ibid., 256.
3. Putnam, *Lighthouses and Lightships*, 18.
4. *Historically Famous Lighthouses*, USCG Publication 232 (Washington: US Government Printing Office, 1986), 21–22.
5. Lilla M. Hawes, "The Memoirs of Charles H. Olmstead, Part VI," *The Georgia Historical Quarterly* 44 (1960): 60–61.
6. *Historically Famous Lighthouses*, 22.
7. Ibid.
8. *Plain Dealer* (Cleveland, OH), 3 September 1886.
9. *Chicago Daily News*, 1 September 1886.

Chapter Nine

1. The terminology used in this era is oftentimes vague, with "beacon" referring to lighthouse and vice versa. It appears that the early "beacons" would refer to smaller, illuminated navigational aids. Information on these early structures is drawn from a report on the Cockspur Island Lighthouse prepared for the National Park Service as part of the Historic American Buildings Survey (undated, but believed to be from 1994).
2. Primary and secondary information on the history of Fort Pulaski and the Union assault April 1862 is available from multiple sources. The details in this account are drawn primarily from "Fort Pulaski, Georgia," by J. H. Clifford in *All Point*, 18/4 (2013); "The Siege of Fort Pulaski: 'You Might As Well Bombard the Rocky Mountains,'" by D. H. McGee in *The Georgia Historical Quarterly*, 79/1 (1995); and "Official Report to the U.S. Engineer Department of the Siege and Reduction of Fort Pulaski, Georgia," by Brig. Gen. Q. A. Gillmore (New York: D. Van Nostrand, 1862).
3. "Fort Pulaski, Georiga," *On Point*, 18/4 (Spring 2013): 28–32.
4. *Savannah* (GA) *Morning News*, 31 January 1904.
5. *Baltimore* (MD) *Sun*, 12 November 1911.
6. *Evening Post* (Charleston, SC), 22 June 1912.
7. *Evening Star* (Washington, DC), 25 June 1911.
8. Ibid., 27 October 1911; *Springfield* (MA) *Daily News*, 4 November 1911.
9. *Baltimore* (MD) *Sun*, 12 November 1911.
10. *Knoxville* (TN) *News-Sentinel*, 24 February 1936.

Chapter Ten

1. From Sapelo Island Lighthouse application for listing on National Register of Historic Places, 17 July 1997.
2. According to informational material displayed at the site.
3. *Savannah* (GA) *Tribune*, 15 October 1898.
4. *Darien* (GA) *Gazette*, October 8, 1898, reported in *Early Days on the Georgia Tidewater*, by Buddy Sullivan. Privately published, revised edition 2018.

Chapter Eleven

1. *Augusta* (GA) *Chronicle*, 29 January 1861.
2. *Macon* (GA) *Telegraph*, 4 February 1861.
3. Ibid.
4. Ibid.

5 *New York Herald*, 2 April 1862.
6 As described in the 1972 National Register of Historic Places nomination form for the St. Simons Lighthouse Keepers Building.
7 "The Ghost of the St. Simons Lighthouse," *Brunswick (GA) News*, 25 May, 2018 (https://www.exploresouthernhistory.com/gastsimons2.html. (accessed 22 March 2020).
8 October 23, 2002, memo on file at the Coastal Georgia Historical Society referencing an interview with Margaret MacPherson, Frederick Osborne's great-granddaughter.
9 *Brunswick (GA) Advertiser*, 6 March 1880.
10 Ibid., 13 March 1880.
11 *Atlanta Journal*, 7 August 1968.
12 From a contemporary Light-House Board report quoted in "St. Simons Island," by Wayne Wheeler, *The Keeper's Log* (Fall 2016).
13 *Brunswick (GA) Daily News*, 23 September 1904.

Chapter Twelve

1 Camilla M. Merts, "The Story of the Little Cumberland Island Lighthouse," vol. 2 of *The Occasional Papers of Island Labs*, privately printed, April 1999.
2 Ibid.
3 Ibid.
4 Tom H. Wells, *The Slave Ship Wanderer* (Athens: University of Georgia Press, 2009), 8.
5 The official records of the lighthouse board indicate that John A. Clubb did not become the lighthouse keeper at Little Cumberland until February 1859, but based on testimony from the trials that followed the *Wanderer* incident, there is little doubt that, in fact, he held that position in late November 1858. The likely explanation is that his appointment was not "official" until the latter date. Further details and discussion are available in Wells's and Merts's publications on the subject.
6 Wells, *Wanderer*, 26.
7 *Cincinnati (OH) Commercial Tribune*, 22 May 1882.
8 Merts, "Little Cumberland Island Lighthouse," 76.
9 Ibid., 77.
10 Camilla M. Merts, "The Grave at Little Cumberland Island, Georgia," *The Occasional Papers of Island Labs*, privately printed, 1/3 (April 1997). Morgan H. Crook Jr., "Excavation and Reburial of Charles R. Farnham on Little Cumberland Island, Georgia," *The Occasional Papers of Island Labs*, privately printed, 1/4 (April 1997).

INDEX

A
acetylene, 47-51
Aquaman, 10
Argand lamp, 36-37
Argand, Ami, 35-37

B
Bartholdi, Auguste, 48-49
Beauregard, PTG, 61
Bishop Rock lighthouse, 29-32
Brunswick, city of, 147
Bureau of Lighthouses, 63, 65-66, 85

C
Cassini, Giovanni, 19
Clubb, James A., 171-172
Cluskey, Charles B., 152-153
Cockspur lighthouses, 119-131; information for visitors, 130-131
Coffin, Howard, 133
colonial lighthouses, 53-55
compass, magnetic, 13
Confederate States Lighthouse Bureau, 72-77
Cordouan lighthouse, 27, 45
Cumberland Island lighthouse, 165-169

D
Dalén, Gustaf, 47
Darling, Grace, 92-96
de Weldon, Felix, 130
Douglass, James N., 29-32
Dover lighthouse, 16-17

E
earthquake of 31 August 1886, 114, 129, 157, 175
Eddystone lighthouses, 28-32, 34
Endicott Era fortifications, 144, 176

F
Faraday, Michael, 47
Farnham, Charles R., 173-175, 179
Farrand, Ebenezer, 75
Fort Pulaski, 81, 103, 109-110, 121-126
Fresnel lenses, types, 44-45
Fresnel, Augustin-Jean, 41-45
Fresnel, Léonor, 45

G
Galilei, Galileo, 19
Gaynor, Hazel, 10
Global Positioning System (GPS), 67
Gould, James, 148, 150

H
Halley, Edmond, 19
Harrison, John, 21-22, 67
Hastings, Joseph, 165
Hogarth, William, 20-21
Huygens, Christiaan, 19, 42

J
James, P. D., 10

L
Lee, Robert E., 120, 123
Lewis, Idawalley Zoradia ("Ida"), 96-100, 129
Lewis, IPW, 60
Lewis, Winslow, contracts with US government, 56-61; lamp 39-41
Lighthouse Board, 50, 58, 61-63
lighthouse keepers, 83-100; duties 83, 85-87; heroes, 90-92; heroines, 92-100; personal and family life, 83-84, 88-90; salary, 87-88; uniforms 86-87
Lighthouse(s), ancient, 14-17; construction, 25-32; differentiation between, 50-51; during American Civil War, 69-82; eras in US, 84-85; Golden Age, 24; illumination, 33-52; purpose, 8-9; symbolism, 10-11; wave-swept, 26-27
Lincoln, Abraham, 72-73
Little Cumberland Island lighthouse, 164-180; advice to would-be visitors, 180; decommissioned, 1915; preservation, 176-180
Longitude Act, 20-22
longitude, 18-22
LORAN, 51

M
Martus, Florence, 127-130
Meade, George, 61

N
navigational advances in 20th century, 66-67
NAVSAT, 67
Newton, Isaac, 19, 42
Norris, John S, 121

O
Oglethorpe, James, 103, 107, 119, 123, 145-147
Olmstead, Charles, 109-110, 124-126

P
Pharology, 15
Pharos of Alexandria, 15-16
Pharos of Ostia, 16
Pleasonton, Stephen, 57-61
Poe, Edgar Allen, 10
Price, Eugenia, 10
Putnam, George R., 62-65

R
Reynolds, Richard J., 133
Rogers, Thomas, 39, 41, 43

S
Sapelo lighthouses, 133-144; constructed by Winslow Lewis 1820, 135-136; decommissioned 1905, 140; information for visitors, 142-144; new lighthouse 1905-1933, 141-141; range light, 137; restored 1998
Semmes, Raphael, 61, 71-72, 75
Shovell, Sir Cloudesley, 20

INDEX

Soleil, François, 43, 161
Spalding, Thomas, 133
sperm oil, 40, 47
St. Simons lighthouses, 145-162; 1810 lighthouse, 148-151; 1872 lighthouse, 152-155; during Civil War, 150-152; information for visitors, 162; Keeper's Dwelling Museum,159-160; killing of lighthouse keeper,156-157
Stature of Liberty, 48-50
Stevenson, Robert Louis, 10

T

transatlantic trade, 22-23
triangular trade, 22-23
Trinity House, 9,26, 47, 92
Tybee lighthouses, 103-118; during Civil War, 108-110; Fort Screven, 114; information for visitors, 117-118

U

United States Lighthouse Establishment (USLHE), 55-56
United States Lighthouse Service (USLHS), 63

V

Verne, Jules, 10
Vespucci, Amerigo, 19

W

Wanderer slave ship, 169-172
Waving Girl. *See* Martus, Florence
Werner, Johannes, 19
Wesley, John, 119
Winfield Scott's "Anaconda Plan," 73-74
Winstanley, Henry, 9, 28
Wolfe, Virginia, 10

ACKNOWLEDGMENTS

Writing a book is always a challenge, a task made easier by those individuals and institutions offering assistance, insight, advice, and support. This work on Georgia lighthouses was greatly enhanced through the kindness of many to whom I owe a debt of gratitude. Sarah Jones, executive director of the Tybee Island Historical Society, and Mimi Rogers, curator of the Coastal Georgia Historical Society, both offered their time and valuable resource material that made my research far easier than it might otherwise have been. The online resources of the United States Lighthouse Society, "a nonprofit historical and educational organization dedicated to saving and sharing the rich maritime legacy of American lighthouses and supporting lighthouse preservation throughout the nation," are superb and were a valuable resource and guide in my research.

Perhaps without saying, one cannot write about lighthouses without viewing them personally. I made multiple visits to each of Georgia's lighthouses whenever possible. The staff at the Tybee and St. Simons lights was most helpful and informative. My several-day stay on Sapelo Island was made all the more pleasant by my lodging in Kathy Hagler's cottage in Hog Hammock, together with the use of her pickup, which allowed me to visit the Sapelo lighthouse at odd hours. Peter and Miriam Lukken entertained me royally while I was there. Members of the Little Cumberland Island Homes Association were gracious in inviting me to visit their island, in addition to providing unique photographs and documents pertaining to the Little Cumberland light. I had the pleasure of touring the lighthouse with several members of LCIHA's Lighthouse Committee. Michele

Hunter was especially kind in spending the day with me on the island, treating me to a comprehensive tour of its natural beauty.

As with my previous five books, I appreciate the support and encouragement offered by the dedicated staff of my publisher, Mercer University Press. And, as before, my muse Laura Ashley was frequently with me, relentlessly prodding me on through the long and often tedious writing process.